WITHDRAWN
NDSU

RHETORIC OF MACHINE AESTHETICS

Barry Brummett

Westport, Connecticut
London

Library of Congress Cataloging-in-Publication Data

Brummett, Barry, 1951–
    Rhetoric of machine aesthetics / Barry Brummett.
      p.   cm.
    Includes bibliographical references and index.
    ISBN 0–275–96644–5 (alk. paper)
      1. Aesthetics, Modern—20th century. 2. Machinery—Aesthetic aspects. 3. Culture—Semiotic models. 4. Popular culture.
I. Title.
BH301.M25B78   1999
111′.85—dc21   99–18012

British Library Cataloguing in Publication Data is available.

Copyright © 1999 by Barry Brummett

All rights reserved. No portion of this book may be reproduced, by any process or technique, without the express written consent of the publisher.

Library of Congress Catalog Card Number: 99–18012
ISBN: 0–275–96644–5

First published in 1999

Praeger Publishers, 88 Post Road West, Westport, CT 06881
An imprint of Greenwood Publishing Group, Inc.
www.praeger.com

Printed in the United States of America

The paper used in this book complies with the
Permanent Paper Standard issued by the National
Information Standards Organization (Z39.48–1984).

10 9 8 7 6 5 4 3 2 1

For Detine Bowers

Contents

Acknowledgments		ix
Chapter One	Aesthetics, Machines, and Rhetoric	1
Chapter Two	Mechtech: Classical Machine Aesthetics	29
Chapter Three	Electrotech: High Technology Machine Aesthetics	57
Chapter Four	Chaotech: Aesthetics of the Decayed Machine	89
Chapter Five	Simulations and Machine Aesthetics in *Brazil*	115
Works Cited		139
Index		147

Acknowledgments

If it takes a village to raise a child, surely it takes an academic community to bring up a book in the way it ought to go. I have an excellent and supportive academic community in the Department of Communication at the University of Wisconsin–Milwaukee, and I thank all of my colleagues and students here for their support. I particularly thank Kathryn Olson and Detine Bowers who read early drafts of some chapters.

I also wish to thank the University of Minnesota and Professor Robert L. Scott, Macalester College and Professor Roger Mosvick, and the Center for Twentieth Century Studies here at UW–Milwaukee—all three invited me to lecture on the rhetoric of machine aesthetics during early stages of this project, and the feedback of faculty and students was quite helpful. Finally, thanks to a wife and two daughters who are tolerant of "Daddy's new toys," acquired from time to time (the latest is a cool, tiny cel phone), that express his/my love of machine aesthetics.

CHAPTER ONE

Aesthetics, Machines, and Rhetoric

> Great technology is beautiful technology.
> —David Gelerntner, 129

On a busy street just north of the inner city of Milwaukee is a new strip mall. Its "look" makes it stand out from among the usual stretch of office buildings and Burger Kings along that street. The stores are the expected mix of Walgreens, Office Depot, and so forth. But the exterior of each store is a smooth, geometric, metallic gray punctuated by enormous red steel girders emerging like a scaffolding from the exterior, industrial flying buttresses anchoring buildings to the parking lot. The whole mall looks like a factory full of machines that might be set in motion by turning a switch, an enormous engine of commerce.

As I write this manuscript I find I am surrounded by rounded, flowing shapes, all of the same neutral light beige: my computer and its monitor, the printer standing next to it, a cordless telephone on a nearby shelf, and an electric spaceheater composed of a series of identical slim, standing, rounded rectangles with simple controls and switches on the face of the last pillar. Each quiet form hides a potent electric power to create, to communicate, to command my environment. They work, I know not how.

On the grounds of a state park an hour from my home is an old quarry lake. The turn of the century mining machinery dug down, down, down until it hit a spring. By then the lode was played out and the machinery too large and bothersome to remove, so the pieces were left there to drown as the waters rose. Spring-fed, the lake is startlingly clear, objects within it visible to a dreamlike depth. You can sometimes make out the rusted hulks moldering at the bottom among the fish.

As these three examples show, we live in a world of machines and technology, and in a world that looks, feels, and sounds like machines and technology. Since the complicated clocks of the Middle Ages and da Vinci's Renaissance visions of flying machines, people have sought out the sensory experience of the machine, and sometimes, of the machine in ruins. We often find those sensations where there is no actual machine: stylish black clothing that has more metal parts to it than it needs to for functional purposes, streamlined furniture that will never move through the air, and rehabilitated factories and warehouses turned into condos for the urban elite yet retaining seized-up freight elevators, fire doors, and exposed, rusting iron beams reminiscent of an industrial past.

Art may be sought out "for art's sake," but an aesthetic experience usually has wider effects. People enjoy the sensory experience of machines, but that enjoyment can lead to further attitudes, actions, and commitments. A preference for a brushed stainless steel kitchen is aesthetic, but it is not *only* aesthetic. It may support a commitment to efficiency and regimentation at home as well as in the factory.

A friend of mine claims only sensory enjoyment from his wandering the industrial wasteland of abandoned factories in the Menominee River Valley that winds through the center of Milwaukee. But I think he emerges from that aesthetic experience with a grim satisfaction at the ongoing decay of an industrial infrastructure that makes his working life unhappy and that consumed his working class family. I suspect that those who prominently display beepers and cel phones want to say something to others and perhaps to themselves through those devices about their status and future prospects in life. They are *using* the devices as *signs* in the hopes of achieving specific effects.

Beyond our sensory enjoyment of machines lie the uses and effects of those aesthetic experiences. Those effects are *rhetorical*: we influence others, and they influence us, in many ways. This book studies the ways in which the look, feel, sound, and other sensations of machines, whether or not in real machines, is part of the machinery of rhetorical influence in contemporary culture.

What rhetorical potency lies within the computer to make it such a scary character in the films *2001: A Space Odyssey* or *Wargames*, and to what political or social effect? How does the heavy use of the heavy machinery of buses, elevators, and firearms contribute to the excitement and power of the movie *Speed*? Why is the anti-fascist film *Brazil* full of broken and obsolete machinery, and why do the characters of the post-apocalyptic *Mad Max* movies wear clothing made from gears, fenders, and springs— to what *rhetorical* effect are these machines and images of machines put?

The rhetorical potency of the machine and its representations goes far beyond the cinema, of course. That potency is all around us. The rhetorical effects of machine aesthetics are embodied in the meanings and significations facilitated by different aesthetic experiences. In other words, an aesthetic experience is meaningful, and that meaning lends itself to rhetorical manipulation. We need a vocabulary for understanding the potential that machines and machine images have for meaning, and a sense of how those meanings can be connected and channeled into rhetorical purposes. To achieve those understandings is the goal of this book. This book develops a *rhetoric of machine aesthetics*. Specifically, this book addresses these research questions:

- What is a rhetoric of machine aesthetics and why would such a rhetoric be important?
- What are some types and categories of machine aesthetics, and how might such categories contain the potential for rhetorical influence?
- How does the rhetoric of machine aesthetics work in specific texts?

This first chapter develops a working understanding of three key terms: *aesthetics*, *machines*, and *rhetoric*. The many meanings of aesthetics are explored, and an approach to aesthetics based on the works of John Dewey and Kenneth Burke, among others, is developed. The nature of machines and machine images as aesthetic experiences is explained, and then the chapter discusses the ways in which those aesthetics are linked to rhetorical uses. Efforts of other scholars to identify the rhetorical potential of machine aesthetics are reviewed. A scheme of the categories of machine aesthetics is then proposed.

Most of the book is organized around the three types of machine aesthetics introduced in this first chapter. Chapters two, three, and four each examine one type of machine aesthetic. Classical machine aesthetics, here called *mechtech*, is the appeal of metal, gears, pistons—industrial engines. High technology machine aesthetics, here called *electrotech*, is the appeal of computers, personal organizers, keyboards—the electric. The aesthetics of the decayed machine is called *chaotech*, exemplified in the dark beauty of rust belt factories collapsing around silent megahorsepower diesel engines. The three major types of machine aesthetics are compared on a scheme of categories that constitute a grammar of their rhetorical influence. In other words, the potential of each aesthetic for influence, the range of meanings carried in their signs, is explained. Finally, chapter five is an extended critical study that applies the theoretical principles developed here to the movie *Brazil* (Gilliam), to explain how the rhetoric of that film makes use of some machine aesthetics. I will argue there that the central themes of *Brazil* can best be understood by considering the presence of all three types of machine aesthetics in its plot and images.

Let us turn now to the three key terms of this book: *aesthetics, machines,* and *rhetoric*. We need to understand how each term has been used by scholars, designers, engineers, and architects. We will also come to an understanding of how each term will be used in this study.

AESTHETICS

Aesthetics is one of those terms that is widely used yet with little consistency in scholarly discourse, even more so in everyday life. As Francis J. Coleman laments, "We cannot hope for complete agreement as to what is and what is not to be called 'aesthetic' " (8). Alan Holgate agrees: "There is major disagreement concerning the nature of beauty and the feelings it evokes in us. Philosophers, aestheticians, and critics have discussed these problems for over two millennia with little progress" (17).

At least three meanings of *aesthetics* can be identified and explored. Aesthetics can be a *systematic way of thinking* about something, that something usually taken to be art, beauty, or sensory experience. Aesthetics can be a *faculty of appreciation* held or not held by an individual; in this sense one might say "I have an aesthetic appreciation for . . . " some sort of art or object. Aesthetics can be a *property of objects* or experiences themselves, as when one might say "I prefer the aesthetic of the Saab to that of the Volvo."

Aesthetic Systems

As a system of thought, one might agree with Richard Kostelanitz that "*Esthetics* is our word for the philosophy of art, which is to say, abstract thought that defines artistic experience in general terms meant to have comprehensive relevance" (13). Although this statement puts off the question of what is art or "artistic experience," it uses *aesthetics* in the sense of a grand system of thinking. That ambitious sweep of Kostelanitz's formulation might seem too grand for some aestheticians such as Anne Sheppard, who cautions us that no theory "can be extended so as to cover satisfactorily all aesthetic appreciation of nature and of man-made objects." (61). Nevertheless, we can think systematically about something that we define as art, or beauty, or sensory experience; this book is an illustration of that possibility. And we can think systematically without totalizing, being aware of the limits of the system and the discourse that embodies it.

Furthermore, Kostelanitz correctly notes that "esthetics is not exclusively the domain of self-avowed estheticians" (20). Aesthetics dovetails with other disciplines and discourses. Paul Crowther decries the "essentialist" modernist program to separate aesthetics from other disciplines, preferring instead a postmodern flexibility that would join aesthetic thinking with other discourses (vii–x). That stance is consistent with this book's attempt to think *rhetorically* about a particular kind of aesthetics. I believe that an interdisciplinary approach to aesthetics (indeed, to nearly any subject) is facilitated by as simple and straightforward a formulation of key terms as possible, yet an operationalization of terms that permits a catholic transcending of disciplinary boundaries. I intend this book to be based on a "simpler" view of aesthetics described by J. O. Urmson as "the way the object in question looks or presents itself to the other senses" (368). Sheppard seconds that flexible understanding by saying that "aesthetic appreciation may be directed at a variety of natural and man-made objects, perceived by any of the five senses" (57). Now, some aesthetic systems are grounded in the sense of aesthetics as a *faculty* and some are based in aesthetics as a *property*. This book straddles both these second and third senses. To understand why, let us turn to John Dewey's aesthetic system.

Aesthetic Faculties

In an effort to think systematically about aesthetic experience, but not too narrowly, I will ground my systematic aesthetics in part in the work of John Dewey (with some modifications to be taken up later). Dewey's systematic aesthetic features this second sense of *aesthetic*, specifically a faculty that people have for a certain sort of sensory experience. Dewey locates the aesthetic experience in the observing subject rather than in the object of regard. The work of art, he argues, is typically seen as a painting, a building, the performance of a symphony, and so forth. In his view, however, "the actual work of art is what the product does with and in experience" (3); in other words, the difference that objects make in the experiences of people.

Art has no ontological status apart from perception, then, for Dewey: "The poem, or painting, does not operate in the dimensions of correct descriptive statement but in that of experience itself" (85). Art "has esthetic standing only as the work becomes an experience for a human being" (4). In Dewey's view, how people respond to any object or event is what determines whether that is an aesthetic object or event (72–75).

Looking ahead, we can see that this view of what is aesthetic is inherently rhetorical, grounded as it is in art's connection to the subject's experiences, interests, and goals. Dewey believes that art is not art unless it is "linked to the activity" of the experiencer (49). The artistic experience then is one that "the beholder must *create*" (54). Because art does not "lead to" but rather "constitutes" experience, its creative power to constitute that experience one way or another is clearly rhetorical (85).

Locating aesthetics in experience rather than objects widens one's net in capturing that which is aesthetic. Morse Peckham notes that "the Chinese placed interesting rocks in their gardens not in order to make them works of art but because they were works of art as soon as they decided that they were" ("Art," 98). This view of what is art allows aestheticians such as Urmson to argue that "we also derive aesthetic satisfactions from artifacts that are not primarily works of art," (357), and it does not take a big leap of discernment to see that such a stance opens up the world of machinery to aesthetic analysis.

Sheppard agrees with Dewey in arguing that aesthetics should be about "appreciation" rather than about works of art themselves (64). An aesthetic experience is the ability one has to find a special quality in that experience. Michael Kirby gets at this notion in claiming that "the key word in aesthetic theory is not *beauty*, as has been suggested by traditional aesthetics, but *significance*... as a *quality of perception*" (43).

We know we are in the presence of art when we react to it in certain ways. And these modes of reaction are often, if not always, *socially* determined. As Peckham put it, "a work of art, then, is any artifact in the presence of which an individual plays a particular social role. . . . It is any perceptual field which an individual uses as an occasion for performing the role of art perceiver" (97). There are socially determined ways to have an aesthetic experience in the presence of a ballet, ritual scarring, or bull fighting.

Peckham's argument points forward to the rhetorical dimension of aesthetics. If faculties of appreciation are socially determined, then they are rhetorically shaped. An aesthetic experience is grounded in a social context and linked to social roles, all of which are rhetorically manageable.

To think of aesthetics in terms of a faculty of appreciation raises the possibility that not everyone will have the same faculty to appreciate an aesthetic dimension of every experience. We might call this the "leaves me cold" phenomenon; one person has a faculty for appreciating the ballet, or the Three Stooges, or quilting, while another person finds no aesthetic experience in each case. For our purposes, we should note here that not everyone has a faculty with which to appreciate machine aesthetics. Gelernter, for instance, claims widespread resistance to the idea that there is machine beauty (18). Some scholars evince no aesthetic awareness of the machine even when writing about issues such as "creativity" or "design." Arthur H. Burr and John B. Cheatham define a machine as "a combination of mechanisms and other components which transforms, transmits, or utilizes energy, force, or motion for a useful purpose" (1), avoiding any mention of a machine's aesthetic dimension. More telling than that, however, is that their section on "analysis and creativity" considers machines solely in terms of utility, with no aesthetic dimension ascribed to "creativity" at all (3–5). Notice that although he is defining engineering *design*, Morris Asimov thinks of it nonaesthetically as "a purposeful activity directed toward the goal of fulfilling human needs, particularly those which can be met by the technological factors of our culture" (1). Alexander McClintock's exploration of *The Convergence of Machine and Human Nature* has nothing at all to say about the aesthetic dimension of that nature. Jacques Ellul's statement that "the machine is solely, exclusively, technique" seems to find no place for the aesthetic (4). This scholarly inattention to the aesthetic dimension of the machine parallels what is for some a personal lack of that faculty of appreciation.

Aesthetic Properties

The third sense in which one might refer to aesthetics attributes aesthetic properties to objects or experiences themselves. In this sense we refer to "the aesthetics of baseball" or to "the aesthetics of bas-relief." This way of speaking locates the aesthetic experience in "reality" rather than in the perceiving subject, and would thus seem to be at odds with Dewey's aesthetic, which grounds the sense of aesthetics I use in this book. However, I want to use this third meaning of aesthetics as a modification on Dewey, a bending of his thinking that will make him more useful and, I hope, more realistic.

To modify Dewey, I will also use some of the aesthetic ideas of Kenneth Burke. In many major works throughout the twentieth century, Burke developed theories of art and literature that combined aesthetics, rhetoric, philosophy, and cultural history in profound and insightful ways. Key to Burke's work is the idea that the substance of a text, the words and other symbols comprising it, are the location of motives and the source of action. We do what we do because of how we describe what we are doing (*Grammar*, xv). The centrality of the material text, the poem, the essay, the speech, the advertisement, as a generator of attitudes, motives, and action is highlighted in this passage from Burke's *The Philosophy of Literary Form*:

> Now, the work of every writer contains a set of implicit equations. He uses "associational clusters." And you may, by examining his work, find "what goes with what" in these clusters—what kinds of acts and images and personalities and situations go with his notions of heroism, villainy, consolation, despair, etc. And though he be perfectly conscious of the act of writing, conscious of selecting a certain kind of imagery to reinforce a certain kind of mood, etc., he cannot possibly be conscious of the interrelationships among all these equations. Afterwards, by inspecting his work "statistically," we or he may disclose by objective citation the structure of motivation operating here. There is no need to "supply" motives. The interrelationships themselves *are* his motives. For they are his *situation*; and *situation* is but another word for *motives*. The motivation out of which he writes is synonymous with the structural way in which he puts events and values together when he writes; and however consciously he may go about such work, there is a kind of generalization about these interrelations that he could not have been conscious of, since the generalization could be made by the kind of inspection that is possible only *after the completion* of the work. (20)

Note the extent to which the material words on the page *are* the motives of the writer, because she is unaware of her motives until *after* her words are set in place.

Elsewhere, Burke stresses the importance of the *form* taken by a text as a generator of attitude and motive. In *Attitudes Toward History*, Burke argues that the forms underlying standard, recurring "poetic categories" such as tragedy, comedy, the elegy, and so forth serve audiences as ways to respond to real life situations (34–91). Repeatedly, Burke argues for a theory of aesthetics that grounds aesthetic experience in texts and objects. However, he does not argue for a simple determinism. Instead, for Burke the aesthetic characteristics of texts and objects work as parameters setting boundaries within which people may exercise their own sensibilities and aesthetics. It is that sense of parameters that I would like to use to modify Dewey's aesthetics.

By merging Dewey and Burke, I propose this understanding of our first key term: By *aesthetics* I mean meaningful sensory reactions to experience, characteristics of experiences that facilitate some reactions and appreciations over others, an experience of appreciation or pleasure in those sensory reactions (especially if we remember that one can find pleasure in the ugly, disordered, and painful), and a unifying focus or noticing of sensory experience. An aesthetic reaction to an object or experience, I claim, is the product of an interaction between the socially influenced needs, interest, and perceptions of individual subjects and the parameters set for that experience by particular texts and objects.

The movie *Rebecca of Sunnybrook Farm* provides different parameters, different possibilities, for aesthetic experience than does the film *Straw Dogs*. One *might* exercise one's aesthetic faculty for violence by watching the former; one *could* conceivably find an aesthetic reaction of bucolic joy in the latter. But such experiences are unusual and certainly work against the constraints of the text and socially constrained reading strategies of the audience. It is an important part of the rhetoric of each text that certain components were selected for their ability to set parameters for aesthetic reaction. That is not to say that the reactions are determined or forced, only that parameters are set—sometimes broadly, sometimes more narrowly—within which Dewey's perceiving subject's aesthetic reactions are most likely to be formed.

If this latest formulation of aesthetic reaction to a text, modifying Dewey with Burke's textualism, is to succeed, it asks that the theorist or critic explain the parameters and the most likely aesthetic reactions facilitated by them, for a given kind of text or object. If we want to understand the aesthetics of, let us say, a tractor, we need to identify categories of aesthetic signification relevant to tractors in general. Such categories are neither dimensions only of the perceiving subject, as Dewey would seem to hold, nor properties of texts in Burke's view, but a combination of both. *Aesthetic categories delimit the range of aesthetic reactions made probable by the intersection between a socially influenced perceiving subject (or audience) and the textual properties of an object or experience.* A scheme of such categories explains the range of aesthetic reactions one is likely to have given certain kinds of objects and experiences within certain kinds of social contexts.

Schemes of aesthetic categories are nothing new, not even in discussing machine aesthetics. Writing of aesthetics generally, Francis J. Coleman argues that "the aesthetic" is distinguished by four general categories: pleasure, interest, exercise of discrimination and knowledge, and praise and admiration (17–23). Clarifying such a scheme meets the charge made by Urmson, that "the central task of the philosopher of aesthetics is . . . to clarify the principles on which we select the special set of criteria of value that are properly to be counted as relevant to aesthetic judgment or appraisal" (363). A scheme of the categories of machine aesthetics is developed in the next section.

We have discussed three senses of the aesthetic. Aesthetics as systematic thought would then be exemplified by this book, as a way to think in an orderly fashion about that experience. Aesthetics in the sense of a faculty within the perceiving subject and in the sense of characteristics of an aesthetic object or experience were joined as we merged Dewey with Burke's theories of aesthetic texts. It became clear in this section that identifying a scheme of aesthetic categories was an important prelude to explaining the aesthetics of any part of life, including machines. But we have yet to consider aesthetic categories peculiar to the machine. Having arrived at some understanding of the role played by *aesthetics* in this study, we turn now to consider what will be meant by *machines*.

MACHINES

A machine is the sort of object, experience, or text with which this book is concerned. Lewis Mumford makes a useful distinction between machines and tools: "The essential distinction between a machine and a tool lies in the degree of independence in the operation from the skill and motive power of the operator: the tool lends itself to manipulation, the machine to automatic action" (*Technics*, 10). But the distinction is not hard and fast, nor is our usage. We are considering the sorts of aesthetic experiences that people have of machines *or* of objects and texts that are not machines but that exhibit machine-like qualities. By a machine-like quality I mean that the object has qualities that facilitate the experience of machine aesthetics. Therefore, the object *shares* aesthetic qualities with whatever a culture takes to be its "benchmark" machines. A building may thus not be a machine, but if it shares characteristics with paradigmatic machines that facilitate a "machine aesthetic" reaction, then it would fall within the interests of this study. Such a building during an era of streamlined machines might have an exterior of smooth metal and glass; it might be simple and functional in its construction and appearance; it might have curved exterior surfaces, and so forth.

This study is not primarily interested in products of machines per se, or in machine art such as computer graphics, except in so far as such things reproduce the aesthetics of machines—as some of them do. As Lewis Mumford notes, sometimes machines *help* in the creative process (as in, after his era, computer imaging), yet that help is not within the purview of machine aesthetics (*Art*, 62–63). Herbert Read's question, "Can the machine produce a work of art?" is not ours, for we want to consider the machine and the machine-like *as* a work of art (57). Mumford suggests printing as an example of "typical machine art" in that the look of one typeface or another depends on machines (*Art*, 73) rather than caligraphy.

Products of machines may themselves show very few characteristics of machine aesthetics (mass produced teddy bears, for instance), or they may be machines themselves, or they may be designed intentionally so as to feature machine aesthetics. Cheney and Cheney report that Henry Ford regarded the Model T as a mechanical work of art produced by machines that flaunted itself as machinery: "Design will take more advantage of the power of the machine to go beyond what the hand can do, and will give us a whole new art" (28). Paradoxically, a machine might itself have very little in the way of machine aesthetics, if it is decorated or disguised (as were many machines early in the Industrial Age or during Victorian Romanticism; see Read, 23–27). Mere connection with a machine, or being a machine, does not make a thing of interest to this study; it must feature machine aesthetics.

An Aesthetic of Machines

We begin by considering the claims that many have made that there is a "pure beauty" in machines (Florman, 134). This beauty is celebrated, if somewhat ironically, in Huysmans' 1880 novel *A Rebours*: "Does there exist anywhere on this earth, a being conceived in the joys of fornication and born in the throes of motherhood who is more dazzling, and more outstandingly beautiful than the two locomotives recently put into service on the Northern Railroad" (Setzer, 17). A usually ironic Oscar Wilde spoke seriously about machine beauty in declaring, "There is no country in the world where machinery is so beautiful as in America" (in Gelerntner, 3). Mumford summarized the explosion of machine aesthetics in the nineteenth century by saying that

> We did not expect beauty through the machine any more than we expected a higher standard of morality from the laboratory: yet the fact remains that if we seek an authentic sample of a new esthetic or a higher ethic during the nineteenth century it is in technics and science that we will perhaps most easily find them. (*Technics*, 322)

Herbert Read rhapsodized, during the 1930s, that "we have perfected this all-powerful tool, the machine, and could we but use it with instinctive wisdom, the results might exceed in beauty and splendor any monuments of the past" (12). At the same time, Cheney and Cheney sang that "we are at the beginning of a new world of appearances, beautiful with the peculiar beauty of the machine" (xi). Also writing during the 1930s, John Dewey uses examples of aesthetic experience that in general are remarkably centered around machines: "the fire-engine rushing by; the machines excavating enormous holes in the earth; the human-fly climbing the steeple-side; the men perched high in air on girders, throwing and catching red-hot bolts" (5).

More recently, Gelerntner made the sweeping statement that "machine beauty is the driving force behind technology and science" (9). *Harper's* Forum celebrated "technology's power to turn many small, insignificant things into one vast, beautiful thing" (74). One hacker quoted in the Forum says, "I'm learning PASCAL now, and I am constantly amazed that people can spin those prolix recursions into something like PageMaker. It fills me with the kind of awe I reserve for splendors such as the cathedral at Chartres" (84–85)—an explicitly aesthetic reaction.

Machine aesthetic is often asserted in contrast to other sorts of aesthetic, often those grounded in nature. Ann Ferebee notes that "from the Victorian period until the present design style has remained polarized into machine-oriented and nature-oriented modes" (10). Reviewing a recent display of made objects in the Museum of Modern Art, Sidney Perkowitz remarked that "they embodied in metal, wood, and plastic an idea that underlies the best in science and technology—the notion of elegance, of concise and beautiful solutions to problems, an idea expressed equally well in powerful mathematical equations, incisive physical theory, and marvelous devices" (58). Beyond the museum, Perkowitz praises the beauty of hubcaps: "Every city street, for example, houses a fascinating gallery of techno-art—the works of machine aesthetic that grace the wheels of every passing automobile" (59).

Some powerful claims have been made for the existence and effect of machine aesthetics. De Lauretis, Huyssen, and Woodward's claim that "technology shapes the very content and form of the imagination" seems to ground all aesthetics in technology if not in the machine (viii). Le Corbusier claims that a machine aesthetic is the birthright of those of us who are heirs to the Industrial Age, for "every modern man has the mechanical sense" (115). As a result, many people are what Florman called " 'gadget cultists,' who delight in the machine as an aesthetically attractive object" (129).

Numerous schools of architecture and design have felt that part of their mission was to create a machine aesthetic in the public or among artists if it was not there to begin with. Walter Gropius argued that for the Bauhaus, "Our ambition was to rouse the creative artist from his other-worldliness and reintegrate him into the workaday world of realities" (89–90). This missionary zeal to spread machine aesthetics across a culture was of course extended to consumer objects beyond machines themselves. W. H. Mayall noted that a consumer "may be influenced favorably by a machine [appearance] which relates with current standards in the appearance of, say, architecture and consumer goods" (25). As a result of the aestheticization of both machines and ordinary products, "there is less to choose, stylistically, between the highly technical and the ordinary domestic product" (Hancock, 272).

As a result of those apostles of machinery in marketing, the aesthetic of machines informs objects and experiences in a culture beyond just the domain of technology. In one of his most famous aphorisms, Le Corbusier claims that "a house is a machine for living in.... An armchair is a machine for sitting in" (89) and so forth, which gave rise to a discourse in which, as Holgate notes, "absolutely everything had to be described as a machine for doing something" (203). Gropius felt that machine-made products should bear the aesthetic mark of machines: "It is to its intrinsic particularity that each different type of machine owes the 'genuine stamp' and 'individual beauty' of its products." In contrast, "senseless imitation of hand-made goods by machinery infallibly bears the mark of a makeshift substitute" (53). Of course, to ask that machine-made products express a machine aesthetic would extend that aesthetic to an enormous sector of any industrial culture. One would then realize the vision of Louis Sullivan, a nineteenth-century prophet of machine aesthetic, who foresaw "an aesthetic of all things that man contrives in mass for man's uses, with only a difference of degree between airplane and ash tray, streamline train and toy cart, and vanity case and skyscraper" (Cheney and Cheney, 31).

Mayall notes the extent to which functionalism in architecture followed a machine aesthetic, carrying it beyond technology (18). An early disciple of that style, Fernand Léger, was moved to make this decree: "I have thus to introduce to you a new order—the architecture of machinery" (46). Susan Fillin Yeh notes the influence of machine aesthetic in other disciplines, crafts, and arts as well: "An emphasis on a machine derived style as a characteristic of objects is one of the major currents of a 1920s sensibility. The emphasis on design is also a crucial factor in Precisionism—the machine-ist aesthetic appearing in painting" (18). Home decorating was influenced by machine aesthetic, as Eidelberg observes: "Interiors were easily brought into conformity with this new style" between 1935–1965 (73). Henry M. Sayre's interesting work shows how the poet William Carlos Williams used metaphors based on machine aesthetic to describe his own poetry.

Machine aesthetics is sometimes spoken of in the sense of a faculty of appreciation, often described as a faculty held particularly by engineers (Florman, 118–19). The presumption of such a faculty meant that in the 1920s, "engineers were spoken of as artists" (Yeh, 7). In the 1930s, Dewey similarly remarked, "The intelligent mechanic engaged in his job, interested in doing well and finding satisfaction in his handiwork, caring for his materials and tools with genuine affection, is artistically engaged" (5).

That vision of engineering persists, for in the 1990s Samuel C. Florman refers to "the poetic vision of our profession" and claims that "to *become* artists . . . has been the existential pleasure of several thousand engineers" (119–20). He speaks of the engineer's "love of the machine for its intrinsic beauty" (132) and thinks of "an engineering project as the staging of a production at one of the great opera houses of the world" (149). More recently, observers such as Sherry Turkle have noted among scientists "the intensity of their relationships with objects," and she cites an MIT professor who had an aesthetic attraction toward a broken radio in his youth (Romantic, 228). Gregory Benford has called for critical analysis of "the aesthetic value of scientific truth" (225).

In sum, many observers have identified a machine aesthetic. Scholars have shown its extension beyond machines alone. It is time to consider the categories or dimensions of machine aesthetics.

Some Canonical Schemes of Machine Aesthetics

Earlier in this chapter we discussed the need to develop a scheme of aesthetic categories if we are to understand the aesthetic appeal of types or classes of experiences. Now that we have clarified what is meant by a *machine*, it is time to merge that understanding with our discussion of *aesthetics* and work towards the construction of a scheme of categories of machine aesthetics.

Several authors have offered, with greater or lesser degrees of specificity, canonical schemes of the categories of machine aesthetics. Here we review some of them. Cheney and Cheney laid out these grand principles of machine aesthetics:

> 1. Materials are used honestly, each in accordance with its own intrinsic properties, its adaptation to machine processes, and its appearance values. . . . 2. Simplicity is observed in the number and kinds of materials employed, and in the form given to the object, in keeping with the requirements of mass production. 3. Functional expressiveness is the artist's foundation. It is insistence upon engineering integrity as the starting point. (14–15)

Mumford also offers a list of "the machine canons" governing an aesthetic (*Technics*, 352): "The elegance of a mathematical equation, the inevitability of a series of physical inter-relations, the naked quality of the material itself, the tight logic of the whole" (350). Yeh offers this canonical scheme of 1920s machine aesthetic: "functional in design, with a fine finish and a Euclidean, geometric perfection, and neatness" (15).

Alan Holgate summarized the engineer's aesthetic principles as "elegance, clean and clear outlines, and unity of form and finish" (1). W. H. Mayall names "figure-ground" considerations as a paramount aesthetic category for machine design, governing many other categories such as color (35–36). Later, he identifies a second important aesthetic dimension: "the [machine] should appear to be an orderly whole" (63). And two more categories Mayall adds to his scheme are "simplicity of outward form" (70) and "the need for unity of overall form" (80). Some thinkers have suggested schemes specifically of high tech, or what we are here calling electrotech, aesthetics. Turkle argues that the science of artificial intelligence is organized around two "aesthetics" of pattern-recognition versus information processing (Romantic, 236).

Within canonical schemes of aesthetics, some thinkers highlight some particular categories over others in explaining aesthetic reactions, as in Anne Sheppard's claim that "appreciation of formal arrangement also plays an important part in our aesthetic response to man-made objects which are not works of art: it is the form of the elegant mathematical proof or computer programme that we pick out for aesthetic approval" (59). Dewey seems to agree in stressing the importance of *order* in aesthetic values (14–15, 38). However, a "super" category of machine aesthetics is that of *functionalism*: a feeling that form (or the appearance and feel of the machine) should follow function (or what it is that the machine was designed to do). Let us briefly examine this widely recognized dimension of machine aesthetics.

Form and Function

Functionalism is the idea that "the essence of machine art is the expression of a function" (Mumford, *Art*, 71). Functionalism is defined by George H. Marcus as the belief that "if an object is made to function well, it will by definition be beautiful" (12). That which is functional is beautiful, in this view. What reveals or features the functioning of a machine will then be essential to its aesthetic. Functionalism is expressed in the idea that "all form dictated by economic and technological considerations has potential for a more robust form of beauty. It could have more forceful qualities, such as the air of purposiveness and efficiency attributed to many machines" (Holgate, 244). Functionalism as a principle of both industrial design and machine construction has been powerfully influential in the twentieth century. Jacques Guillerme observed that "for over a century industrial production has borne the imprint of stylization . . . adapted to the intertwined requirements of machining and economic calculations" (53).

The idea of functionalism was embraced but altered somewhat by many proponents of machine aesthetics in the nineteenth century who intentionally designed machines to ensure that their form, or appearance, suggested their function. This was true even when, paradoxically, there was no practical reason to do so. The aesthetic of functionalism thus became more important than functional qualities themselves. The heyday of functionalism was likely the last half of the nineteenth and the first half of the twentieth century. Then, Mumford argued that "the American ax, the American clock, the clipper ship—in every line of these utilities and machines necessity of function played a determining part.... For what was beauty? 'The promise of function' " (*Art*, 117).

As functionalist aesthetic came to be more important than function, some argued that the link between form and function is broken. Léger argued even as early as 1923 that "we must not conclude ... that perfection of utility necessarily implies the perfection of beauty" (47). Aldersey-Williams argues that aesthetics simply express function *and more*: "In the messy real world, however, form denotes function, but it also *connotes* additional values" (27). Thomas Hine contends that the belief that style should reflect function has failed, for function does not always reveal beauty (62); he claims that the 1950s were an era in which form superceded function in importance (84). Holgate argues for a break between form and function in principle, claiming that "there is much evidence to support the idea that visual beauty exists at some remove from mechanical perfection, even in the most efficiently designed utilitarian structures" (231). An object then would be beautiful if it *appears* to suggest a certain aesthetically pleasing function rather than if it actually reveals its true function. One example might be *streamlining*, an allegedly functional style of the mid-twentieth century; Martin Eidelberg argues that in fact design was paramount in streamlining, for it was "the conscious intention of creating a futuristic style" (72). The streamlined design of stationary objects, Eidelberg notes, was functional only in appearance (74–75).

Thus, Baudrillard argues that with today's stylizing of machines, "the object, at its farthest remove from objective determinants, is completely taken over by the imaginary" (*System*, 113). Baudrilland suggests an interesting twist on functionalist aesthetic in an " 'aesthetic' approach which omits function altogether and exalts the beauty of pure mechanism" (*System*, 114)—in other words, an aesthetic experience of machines for their materiality, their appearance in and of itself, beyond functionalism. Mumford also appears to argue in principle that function follows form, not the other way around, for "the esthetic symbol preceded the practical use" (*Art*, 67).

Brolin goes so far as to argue that modernism's machine aesthetic in fact often worked against *real* efficiency by creating design imperatives that interfered with true function (33–35). Aldersey-Williams agrees in claiming that from the 1920s through the 1940s form and function very often did not match; instead, the perception that they did was paramount for mechanical design (23–24).

Different canonical schemes for machine aesthetics have been proposed. Many agree in featuring such categories as functionality; others dissent. But many such schemes are keyed to one particular type of machine aesthetic—most, to mechtech. (Mechtech tends to feature functionality, for instance.) My purpose here is to suggest a scheme of categories especially suited to all three types of machine aesthetic examined in this book and to facilitate comparisons among them. I turn now to develop a comparative scheme of machine aesthetic categories.

A Scheme of the Categories of Machine Aesthetics

The following chapters will be organized around a scheme of aesthetic categories that explain the potential aesthetic reactions to each of three general types of machine aesthetics. The scheme also provides a basis for comparison of the three types of machine aesthetics: mechtech, electrotech, and chaotech. Our discussion of machine aesthetics will explicate *nine* categories of aesthetic experience, which will be the basis for comparisons among mechtech, electrotech, and chaotech as illustrated in Figure 1. The nine categories are dimensionality, subject, object, production, gender, persona, dominant relationship, erotic, and motivating context. These nine categories are necessarily overlapping; thus, we will not strain to maintain perfect distinctions among them. They are likely not exhaustive; further research will and should reveal other components of machine aesthetics. Finally, these categories are not limited to machine aesthetic. In explaining them, I will also show their relevance to both machine and nonmachine aesthetics.

The first aesthetic category is *dimensionality*. An important part of any aesthetic experience is reaction to the surface and to the depth of an object (or, perhaps somewhat less obviously, of an experience). That oil painting has greater physical depth than does watercolor, for instance, is an important part of our reactions to each medium. Sculpture in turn has richer dimensionality than does either oil or watercolor, in general. A lily pond is entrancing both for its surface and for the depths visible beyond that surface. A forest in summer presents a skin of green that gives way to an intriguing perception of depth as fall brings different colors. Machines, too, ground the interplay (or absence) of depth and surface in facilitating an experience of aesthetic dimensionality.

Three aesthetic categories are linked: Experiencing the aesthetic object as *subject*, as *object* or *commodity*, and as an instrument of *production*. To use the aesthetic object or participate in an action personally entails a special kind of aesthetic reaction that is distinct from seeing it

Figure 1
Some General Categories of Machine Aesthetics

Dimensions	Mechtech	Electrotech	Chaotech
Dimensionality	Depth/surface	Through the looking glass	The breached hull
As subject	Rhythmic, patterned	Male (see gender), will, instantaneity	Nostalgia
As commodity/object	Geometric, metal, danger, oiled	Female (see gender), streamlined, mystery, secret	Corruption, dirt, rust
Aesthetics of production	Efficiency, uniformity, perfection	Complex-made-easy, mystery revealed, secret discovered	Gridlock, dysfunction
Gender	Male	Androgynous	The toon
Persona	Warrior	Magician	Thief
Dominant relationship	Violence	Seduction	Rape/impotence
Exotic	Body	Mind	Death
Motivating Context	The factory	The wild west	Rust never sleeps: imperatives

as an object. To dance in a ballet invokes one kind of aesthetic linked to but distinct from watching the ballet as an object. The subject aesthetic experience is one of identification, of investment of the self with the object or action; it becomes us, and we become part of it. The object experience distances the aesthetic object or action. It becomes something to be regarded, to be interacted with, perhaps opposed and rejected, perhaps accepted, but always as something "out there." Different from these two, but conceptually related, is to experience aesthetically in terms of what an object or action *produces*, whether that be a motion, an experience, or other objects. This aesthetic category is particularly appropriate to machine aesthetics and may be somewhat less relevant to other sorts of aesthetics. It is the appreciation of what a table saw does as it cuts the wood, producing smooth planes and precise angles again and again in perfect order. That aesthetic is somewhat different from regarding the saw as an object in and of itself, which may entail some fear and alarm for its violent power. And both those aesthetics are somewhat different from the feeling of operating a table saw oneself, feeling the vibrations, the resistance and release of steel roaring through wood, the rise and fall of energy under the control of our fingers.

The extent to which aesthetics is informed by *class* is largely located within these three categories. Working class people are more likely than are upper class people to identify with machines as subjects and to admire their productive capacity or to resent them for their role in a working life of toil, danger, and uncertainty. This is because working class people are more likely to be those actually using drill presses or X-ray machines and to see the actual processes of production. Working class people are most likely to be affected by the decay of machines, to feel nostalgia at the sight of a tractor one once used now rusting and broken, to remember where that piece of metal now adorning an avant garde sculpture actually fits on a generator. Working class aesthetics for these reasons likely skew more towards experiences of monster truck rallies and upper class aesthetics likely skew more towards attending science and technology expositions. In either case, thinking about the subject, object, and production categories of the experience can help to locate the classist aesthetic dimension.

Aesthetics are *gendered*. Not only does it make sense to inquire as to the gender of the aesthetics of a tea room or a sports bar, nearly everyone could instantly identify those aesthetics as female and male respectively. Hunting has an aesthetic that most people would experience as male, while the paintings of Mary Cassatt likely "feel" female to many people. A go-cart is a male machine, an electronic organizer is female. Here especially we see the effects of the social creation of aesthetic faculties within perceiving subjects, in conjunction with socially influenced cultural creations. We are made into "typical males" at the same time that televised professional football is constructed as a type of text disposed to facilitate a "male" aesthetic experience. We will see that machine aesthetics are also gendered in ways that both respect and confound the usual categories of male and female.

It may be entirely whimsical to suggest that an aesthetic experience implies a *persona*, but I believe that at least for machine aesthetics this is true. It may well be an important aesthetic category beyond our scope here. If demolition derbies support a male aesthetic, is it too much of a stretch to argue that they also suggest the persona of the rampaging ogre? Can we argue that the aesthetics of Oscar Wilde's plays and poetry bespoke the image of a fair youth holding a lily? The films of Oliver Stone, so many of which seem to feature soldiers, may be based on the aesthetic persona of a soldier regardless of their subject matter. I will argue that each type of machine aesthetic suggests a fairly distinctive persona. This truth has been recognized in popular culture and its myths. Mechtech has been expressed in the image of John Henry, electrotech in the image of the greasy hacker nerd. We will consider the more formal, abstract persona generated by each kind of machine aesthetic.

Objects and experiences facilitate aesthetic reactions that seem to suggest a *dominant relationship*. Think of how the different aesthetics of football and baseball may be distinguished at least in part by an implication growing out of their aesthetics of how people should and do relate: in direct confrontation or askance, at a distance? Through honest conflict or deceit and trickery? En masse or by way of individual heroics? Think of how the aesthetics of a small, white country Protestant church imply a gathering together of people in quiet intimacy, while the aesthetics of a sprawling European cathedral suggest the scattered yet unified, complicated relationships of the growing urban centers that were arising as they were built. Machines, too, suggest ways of living and relating with others. A personal stereo implies a different mode of relationship from an assembly line. We might think of the ways in which the combination of high technology and crumbling urban infrastructure in *Blade Runner* suggests the movie's isolated, bleak, commodified relationships among loners.

Popular usage has unfortunately captured the term *erotic* and turned it almost entirely into a sexual word. But as Gilles DeLeuze and Felix Guattari, and Herbert Marcuse, among others, explain it, *erotic* more appropriately denotes the larger category of *an order of desire*. The term may thus have as much of Marx as of Freud, of the economic as of the sexual, about it. An academic department may be based upon an erotic of publication, or of choice teaching assignments, or of scholarly awards, in this sense. A family may be driven by an erotic of obligation, incurring and repaying debts of guilt. Aesthetic reactions are clearly based on an order of desire, or erotic. To experience pleasure, to pay attention, to focus, to have any of these hallmarks of the aesthetic is to invoke an order of desire. We have pleasure because we desire some things but not others; we pay attention to what is sanctioned by our order of desire.

Machines are often turned to as instruments of desire, to achieve some effect. We want to *escape* at the end of a bad day at the office, and so we hit the freeway in our powerful automobiles. We want to see which messages we have received, so we use the computer to check our e-mail. We want a well dug, and so hire a company with a drill. Machines have an erotic, they embody an order of desire, and that order will be one of the categories of aesthetic experience of machines.

A final aesthetic category is that of a *motivating context*. Just as an implied persona and a mode of relationship are aesthetic categories, so is an implied context within which personae act and relationships are formed. Part of the aesthetics of individual paintings within frames, done on the human scale, is precisely the sort of museum that has evolved as a site within which to view such paintings. Within that context, people may move from one engagement to another, each absorbing yet disconnected, within the neutral and disconnected space of the art museum. Context may also be a kind of experience as well as a space. Part of the aesthetics of the martial arts is the clearly suggested motivating context of war.

Different kinds of machines and the ways in which they are experienced also suggest motivating contexts. Motivating contexts for machine aesthetics are here expressed as metaphors for understanding the social and cultural milieu most strongly suggested by each type. I will argue, for instance, that mechtech most "fits" with the motivating context of a factory; understanding that context helps us to understand the social implications of the aesthetic. As we will see, this aesthetic category is one of the most unifying and integrative dimensions of machine aesthetics.

We have now considered two of the key terms in this study, *aesthetics* and *machines*. The presence of the third key term, *rhetoric*, has surely been apparent in the discussion so far. Let us now turn to consider explicitly what is meant by that term and how it connects to machine aesthetics.

RHETORIC

Rhetoric has been variously defined throughout history in ways such as pandering (Plato), as the faculty of discovering the available means of persuasion (Aristotle), and as the process of seeking identification and common ground among people (Burke, *Rhetoric*). This study is based on a loose understanding of rhetoric as persuasion, as the creation and management of meaning, as influence through the use of signs. As such, our understanding of rhetoric will be quite broad, delimited by the function of persuasion and influence it carries out rather than by any particular manifestation it may take. While the most traditional manifestations of rhetoric are in the form of public speaking or other sorts of expositional, argumentative forms, here we focus on what signs within texts do rather than on the sort of signs or texts they are. Whatever creates and manages meaning, whatever influences people, whatever results in power gained or lost through signs and symbols is therefore considered to have a rhetorical dimension.

Rhetoric and poetics (or aesthetics) have had a long and close relationship. Howell charts the ways in which rhetoric and aesthetics have intertwined throughout history, both discourses having to do with how people are moved. Likewise, rhetoric and poetics have had a dual nature consisting of both the *study* of a practice and the *practice itself*. Their linked problematics are expressed by Dewey: "Because the objects of art are expressive, they communicate" (104). Donald C. Bryant notes that rhetoric and poetic have always had a close relationship, with the balance in that relationship shifting over time; he observes that ancient Greece and Rome saw the "virtual possession of poetic by rhetoric" (1). The situation has been reversed in other times, such as the late Middle Ages and Renaissance.

An important link between rhetoric and aesthetics has been what McKeon would call their *architectonic* properties: the ability of both to *order* thoughts and experience comprehensively. Steve Whitson and John Poulakos note this shared propensity to order in the work of Nietzsche: "For Nietzsche, rhetoric is not an epistemological undertaking but rather part of a greater artistic act—the act of ordering the chaos of life" (136). For that reason Whitson and Poulakos would merge the two discourses on these grounds:

> Understood aesthetically, rhetoric allows people to suspend willingly their disbelief and be exposed to a world other or seemingly better than the one with which they are familiar, all too familiar. That is why the rhetorical art asks not for dialectically secured truths but for linguistic images that satisfy the perceptual appetites or aesthetic cravings of audiences. (138)

Reason is an ordering process, and reason is often taken to be a dimension of rhetoric. Like Whitson and Poulakos, Dewey sees the aesthetic and the "intellectual," or reason, as inseparable: "Esthetic cannot be sharply marked off from intellectual experience since the latter must bear an esthetic stamp to be itself complete" (38); in other words, one cannot reason unless one does so in the ordered way shared by aesthetics.

When we speak of a rhetoric of any sort of aesthetics, we are speaking of a natural extension or outcome of a continuous discourse. People have aesthetic reactions to experience, and discourse then uses the meanings created by those reactions to create further commitments, thoughts, and reactions. If there is a *rhetoric of machine aesthetics*, it will be *the discursive uses of aesthetic experiences of machines*. Such a rhetoric may be planned, anticipated, and strategized because of the parameters inherent in machine aesthetics, explained in our scheme of categories above. Or such a rhetoric may be unintended outcomes of the conjunction of texts, contexts, and audience reading strategies. A film, for instance, might use the audience's proclivities to read chaotech in certain ways, including the parameters for reading inherent in chaotech texts, to move an audience to accept further feelings, attitudes, dispositions, or actions. Hence the ability of films such as *Escape From New York* or *Escape From Los Angeles* to create anti-governmental, conspiracy-fearing, subversive reactions in an audience out of aesthetic reactions to ruined cities and the broken people who inhabit them.

Technological Rhetoric

Machines, images of machines, discourse about machines, and aesthetic dimensions of machines have long been understood to contain potential rhetorical power. That impact is sometimes expressed in terms of political or social impact by those who are not speaking from a rhetorical vocabulary. The machine as object, image, or subject of discourse is used to manage meanings and commitments in a society. David E. Nye, for instance, in his study of rural electrification argues that technology (and especially electricity) *always* has social and political impact (x); thus, one might argue that it always has a rhetorical dimension.

Machines and their aesthetics pick up connotations that are rhetorically usable. Machines may take on moral or ethical weight, for instance. Holgate tells us that "moral values are often applied in the appreciation of built form. They may be expressed as a demand for 'honesty' in design: an insistence that appearances should reflect the realities of construction, or in a strong distaste for ornament and perceived extravagance" (5). Others have observed that part of what designed or stylized objects such as machines reflect is "social aspirations" of the public (Greenhalgh, 21). This idea of social desire encoded in style adds a rhetorical dimension to machine aesthetics.

The possibility of a rhetoric of machine aesthetics is immanent in many critical and scholarly works. Baudrillard argued that the technological development of objects determined social and psychological reactions to them (*System*, 5), so that how we think about experience is influenced by our machines and ways of using them: "Modes of the imaginary follow modes of technological evolution" (*System*, 118). Istvan Csicsery-Ronay, Jr. agrees in arguing that "our imaginations are increasingly determined by the problems and technical solutions latent in the social application of the given technologies" (27). When thinkers such as Le Corbusier write deterministically that "machines will lead to a new order both of work and of leisure" (95), they locate a power to change society, and thus a rhetorical power, in the machine itself.

Conflicting meanings generated by machines and their aesthetics underlie rhetorical struggle. Like any transcendent symbol, the machine's ability to mean many different things allows it to be used on different sides in rhetorical struggles. Ewen and Ewen mark that polysemy when they observe that "the epoch of the machine is one of hope and horror, ambiguous and confusing" (1); yet it is that very ambiguity that creates the possibility for machines and their aesthetics to be used in rhetorical struggle. Holgate argues that political struggle between "left" and "right" is reflected in the meanings of design: "People whose politics are to the right of centre are often classicists and favour 'redundancy' (such as ornament) in art while those to the left are often functionalists who give priority to efficiency and cost effectiveness" (5). Seemingly in agreement with that idea, Carroll Pursell identifies contrasting democratic and authoritarian "technics" so as to examine the ways in which technology and political stances interact.

Some have argued that a rhetoric of machines is intentional, that machines are designed to bring about cultural results. This notion of strategy is expressed by Florman's contention that "the first and most obvious existential gratification felt by the engineer stems from his desire to change the world" (120). Eidelberg offers as examples of such intentional technological rhetoric the phenomenon of "state gigantism," reflected in the aesthetic appeal of huge dams, power stations, and massive factories (12–15). Nye argues that through such huge mechanical projects, "Society seems to dominate nature by taking control of elemental forces and siphoning them off for human use" (355), a rhetorical strategy bolstering government and capitalism.

But technological rhetoric need not be intentional to have profound effect. A desire simply to increase sales may underlie rhetorical strategies encoded in design and engineering with broader social impact.

W. H. Mayall notes that many products are designed for "social" reasons, often to express certain values, such as tractors that look like automobiles (25–30). Such a tractor will not only have rhetorical effect in its sales but in its impact on how the farmer who buys it thinks about the relationship between work and leisure.

Several authors have suggested more specific ways in which technology, or more specifically machines, have rhetorical power—a power often derived from machine aesthetics. Gary Lee Downey observes that

> throughout American history, a common strategy for producing social change without debating it explicitly has been to delegate the change to technology. This strategy often remains inexplicit because popular theorizing in America imagines technology not as a social phenomenon but as a force external to society that impacts on it and to which society must adapt. (200)

Although Downey does not say so, his interesting observation raises the possibility that technological *aesthetics* are the engine of its rhetorical power "without debating it explicitly." William Washabaugh's interesting work shows that technology can affect aesthetics rhetorically; he studies the impact of electronic recording on traditional flamenco music (62–68).

Another site of technological rhetoric grounded in aesthetics is identified by Ferebee:

> There is a direct ratio between the importance of the modern artifact and the extent to which it is decorated. Scientific instruments . . . lack ornament and are therefore appreciated as sculpture by those who share the contemporary preference for form over pattern. Paradoxically, absence of ornament conveys a message: that modern life is dominated by machinery. (99, 102)

The amount of ornament is, then, a rhetorical statement as to the centrality of the machine in a culture generally.

De Lauretis, Huyssen, and Woodward argue that the effect of separating art and technology has been to deprive art of its affect upon social policy, a separation they decry (vii–viii). Their stance is relevant for our purposes because it merges our three key terms of the *rhetoric* of *machine aesthetics*: it suggests that art, or the aesthetic dimension, is a major way in which a technological rhetoric is *possible*. If there is a technological rhetoric, then de Lauretis et al. claim that "we must begin to understand technology as a relation of the technical and the cultural, as a material and congitive form of social process" (viii).

One important theoretical underpinning to understanding the rhetoric of machines is to think in terms of the possibility of a *system of objects that generates meanings*. Grounded in the work of de Saussure and Peirce, many thinkers have argued that objects in everyday life, including machines, are part of a culture-wide system of signification. In such a system, "we have the interplay of images at work, upon which the meanings we ascribe to products is largely based" (Mayall, 89). Mumford speaks of machines as a systematic code of meanings that facilitates social interaction: "The machine has thus, in its esthetic manifestations, something of the same effect that a conventional code of manners has in social intercourse: it removes the strain of contact and adjustment" (*Technics*, 357).

One of the clearest examples of a theory of systematic meaning is Baudrillard's explication of a " 'spoken' system of objects ... the more or less consistent system of meanings that objects institute" (*System*, 4). One of the most important functions of a system of meaningful objects, according to Baudrillard, is the ability of its meanings to function "as an imaginary resolution of contradictions of every kind" (*System*, 129). That is to say that problems and conflicts within any society are explained and kept under manageable control through rhetorical manipulation of a culture's system of objects. Sanford Kwinter likewise theorizes a system of objects, but his thinking is directed more explicitly to the technological implications of the system:

> Around each and every technical object there may be associated a corresponding complex of habits, methods, gestures, or practices that are not attributes of the object but nonetheless characterize its mode of existence—they relay and generalize these habits, methods, and practices to other levels in the system. Thus it is not in the object that analysis ought to be interested but the complex. (194)

Following Kwinter's advice, we will focus not just on individual objects but on the systems of aesthetics and of the signfication of those aesthetics that make them rhetorical. Each of our three types of machine aesthetic may be thought of as a subsystem of meaningful aesthetic objects.

Holgate has an interesting analysis of how some objects pick up systematic meanings. Using an example explicitly grounded in machine aesthetics, he notes that certain successful objects become paradigms for how similar objects should look. That "look" comes to signify the success and cachet of the original paradigm: "If a Japanese tape deck is by far the most advanced, efficient, and economic in the world, it will not be long before the look of that tape deck becomes 'the look' for tape decks: the 'look' that expresses efficiency" (230). Tape decks and their meanings thus constitute a subset of the subsystem of electrotech objects.

Conclusion

By the *rhetoric of machine aesthetics*, then, I mean the ways in which the categories and dimensions of the aesthetic experience of machines and machine-like signs might be put to use so as to influence meanings, attitudes, values, politics, and the distribution of power. The next three chapters explore three general types of machine aesthetics: (1) Mechtech, or classic machine aesthetics. This is the appeal of industrial machines, of oil and steel. (2) Electrotech, or high technology machine aesthetics. This is the appeal, perhaps increasing at the present time, of computers, lasers, and the electric. (3) Chaotech, or the aesthetics of the decayed machine. This is the appeal, also perhaps increasing at the present time, of the rusted and tumbling down factory, of obsolete machinery, of shorted motherboards and crashed hard drives wrenched out of useless CPUs.

Each type of machine aesthetic will be analyzed according to the nine categories of aesthetic explained in this chapter: dimensionality, subject, object, production, gender, persona, dominant relationship, erotic, and motivating context. Each chapter will conclude with a suggestion of how each aesthetic has been used rhetorically in the past, and what rhetorical potential each may still contain. But in reality, I take an explanation of the categories of each type of machine aesthetic to be *also* an explanation of the rhetorical potential of each type. My purpose here is to explain the *rhetorical repertoire* offered by each kind of aesthetic experience. A *rhetoric* is, in one sense, a description of just such a repertoire: of the tools and strategies available to influence people. So *this* rhetoric, this book, intends to explain the rhetoric of machine aesthetics by explaining the building blocks, the tools, within that repertoire. We will conclude with a critical study that makes use of that repertoire to show how machine aesthetics may be used for persuasive purposes.

CHAPTER TWO

Mechtech: Classical Machine Aesthetics

By *mechtech* I mean a machine aesthetic keyed to gears, clockwork, lawn mowers, revolvers, pistons, hard shiny metal, oiled hot steel, thrumming rhythms, the intricately choreographed blur of a spinning camshaft, and the utilitarian shafts and pipes running through the steel box of a factory. Of the three kinds of machine aesthetics studied in this book, this has certainly been the dominant machine aesthetic of the twentieth century and is still commonly experienced today. What many people think of when they think of a "machine" grounds this aesthetic.

SOME HISTORICAL BACKGROUND

Machines have been part of human experience for centuries, of course, but have grown in dominance and psychological presence since the start of the Industrial Revolution in the early nineteenth century. Western civilization especially has been dominated by machines, in the view of Lewis Mumford (*Technics*, 3–4), and he *means* mechtech machines. Mumford argues vigorously for the centrality of the machine in western thinking, for "no one can hope to achieve any kind of personal integrity in the modern world who is not at home with the machine" (*Art*, 54).

Early industrial machines were often *decorated*, their iron legs and supports worked into twining vines, floral motifs, or claw feet in what Harold Van Doren called "the betrothal of art and industry" (11). Although this art, or design, did not express a machine aesthetic it did in a sense lay a groundwork for it by creating an awareness that machines could be *stylized*. Rhetorical uses of machine aesthetics were evident here, for an artistic machine was presumed to be easier to work at—thus industrialists attempted to pacify their workers.

Decorated machines were more common early in the Industrial Age, and were often an attempt to ameliorate the unaccustomed aesthetics of locomotives and steam engines. Aesthetics in the early nineteenth century, to the extent they were oriented towards machines at all, were usually positioned *against* the perceived grime and ugliness of the factory. Mumford argues that "the very bleakness of the new environment provoked esthetic compensations" (*Technics*, 199). The later Romantic movement in aesthetics preferred to ignore the machine or to view it with horror.

Seeds of twentieth century mechtech aesthetics began to emerge in the latter part of the nineteenth century. The key move involved a growing aesthetic appreciation of machines as machines. Deborah Bright points to photography as an early site of mechtech. She argues that nineteenth century optimism about industrial progress was reflected in a fascination with the machinery of photography, paradoxically when used to capture nature, for "nature and mechanism appeared to spring naturally and autonomously from the same divine principle" (60). That same celebration of industrial power could be found in Stuart Ewen and Elizabeth Ewen's description of

> The "Corliss Engine," a gargantuan device exhibited at the Philadelphia Centennial Exposition (1876). . . . Viewers of the engine were often reduced to tears in its awesome presence, swept away by the apparition that they were witness to the power of the universe, tamed and harnessed by reasonable, enlightened science. (3)

In architecture, Louis Sullivan was identified as a prophet of machine aesthetics in the 1870s by Sheldon Cheney and Martha Cheney, and they note his influence as a model for Frank Lloyd Wright's work (30–31).

But the twentieth century, especially its early years, has been the era of greatest interest in mechtech aesthetics. When people speak about machine aesthetics during the first forty years or more of the twentieth century, they are speaking about mechtech. As Brent C. Brolin notes, "although the delight in machines goes far back in history, the degree to which machines dictated the aesthetic norm in the early years of this century was unique" (52). That norm was mechtech; elecrotech was in its infancy, and any aesthetics of decay were expressed in a Gothic rather than chaotech idiom.

Early in this century, "the engineer became the model for a changed aesthetic sense" (Yeh, 5). Several schools of design or architecture arose early in the twentieth century that have prominently featured mechtech machine aesthetics. Several labels have been applied throughout the century to these aesthetic movements; in this proliferation of names for one school or another we find much overlap but less consistency. What they have in common, as Cheney and Cheney put it, is that "the machine is the foundation fact as well as the shaping tool; is influence and inspiration" (vii). Designers, architects, decorators, and artists alike began to create objects, buildings, and art that had the look and feel of (usually, mechtech) machines, even if they were not machines. Terms such as *cubists, constructivists, Bauhaus, futurists* and so forth arose to describe concentric schools of thought with slightly different philosophies within this overall movement toward machine aesthetics.

Mumford notes that *cubists* or *constructivists* were among the first designers and architects to celebrate machine aesthetics in this century (*Technics*, 335). Cheney and Cheney locate constructivism in Europe in the 1920s and 1930s (36). Ann Ferebee likewise identified cubists, constructivists, and futurists as types of early twentieth century machine aestheticists, explaining that "the Futurists derived the belief that beauty results automatically from perfect mechanical efficiency" (96–97). Alan Holgate claims that the early futurists had "a curious fascination with 'the machine' which was perceived to have mystical qualities, and yet express the 'cold rationality' of its designer" (199).

A celebrated school of design and architecture in Germany was known as the *Bauhaus*, the influence of which "was considerable and long-lived" (de Noblet, 24). A major figure associated with the Bauhaus was Walter Gropius. His work featured, as Elodie Vitale notes, "the integration of art and the machine" and "the integration of art and technology" (158). Looking back over his career, Gropius said, "Believing the machine to be our modern medium of design we sought to come to terms with it" (75).

We should remember from chapter one that the aesthetic category of *functionalism* was often used to pull together the many overlapping systems of thought, design, and architecture featuring mechtech aesthetics. Functionalism was thus seen as yet another school of thought in design. It is defined by George H. Marcus as "the notion that objects made to be used should be simple, honest, and direct; well adapted to their purpose; bare of ornament; standardized, machine-made, and reasonably priced; and expressive of their structure and materials" and he claims that this view "has defined the course of progressive design for most of this century" (*Functionalist*, 9). Functionalism is often proposed as an aesthetic standard, as in David Gelerntner's claim that "machine beauty requires good functioning" (13). Evert Endt and Sabine Grandadam agree, noting that "beauty of form became inextricable from functional considerations and was made accessible to the greatest number. It was the dawn of the age of the mechanical paradise" (31).

Despite the proliferation of terms and rubrics to describe schools of architecture and design, these groups are essentially similar for our purposes here in terms of a shared aesthetic that was powerfully influential early in the twentieth century. Henry M. Sayre notes correctly that "in the 1920s and 1930s a machine aesthetic dominated virtually all modern styles and movements (and those that did not adopt it reacted against it)" (314). Sayre argues that "rhetoric about art and the machine . . . virtually inundated the avant garde journals in this country after World War I" (322). In varying ways, different designers and architects borrowed aesthetics from mechtech machines and from the contexts of machines such as industry and mass production. Stuart Ewen cites the example of Peter Behrens, who "invested all elements of corporate design with the naked, if aestheticized, principles of mass production" (137).

Architecture was one of the major, and perhaps the first, site of mechtech aesthetics throughout the twentieth century (Cheney and Cheney, 141–79). Brolin notes that "in the early twentieth century an architectural revolution took place. All traditional styles were declared null and void. . . . Architects now looked to airplanes, factories and industrial machines for inspiration" (14). Fernand Leger was an early advocate for the new architecture: "Machine Beauty, without artistic intention, is important because of its strictly geometrical and agricultural organization. I have thus to introduce to you a new order—the architecture of machinery" (46)—surely the last time the term "agriculture" was used in such a context!

Probably the most celebrated architect to champion machine aesthetics was the pseudonymous Le Corbusier, who declared in the 1920s that "a great epoch has begun. There exists a new spirit. There exists a mass of work conceived in the new spirit; it is to be met with particularly in industrial production" (9) and more specifically in "machinery, a new factor in human affairs" (84).

Some observers argue that today the classical machine aesthetics of mechtech are no longer dominant in the cultural scene. Martin Eidelberg argues that by the early 1960s an anti-modernist, anti-machine aesthetic emerged in architecture and design (304). Thomas Hine argued that in the 1950s public fascination with the machine look faded: "People at the time seem to have understood that they were living with the benefits of the machine age, but unlike advocates of modern design, they saw no need to celebrate the industrial process" through intentional stylization of objects and buildings in a machine aesthetic (12). The simplicity and clarity of mechtech, still found in public aesthetics, Hine argues, was replaced by "enrichment" and excess in aesthetics in private homes (79). Marcus perhaps overstates the matter in claiming that "by the late 1970s functionalism was a sinking ship" (152). It would be more accurate to say that mechtech no longer wields the aesthetic dominance that it once did, but that it still occupies a strong niche in the culture.

One useful model for thinking about today's mechtech presence is given by Marcus, who argues that classical mechtech has "become synonymous with what is modern, and its antithesis—ambiguity, ornamentation, individuality, and complexity—with what is postmodern" (9, 153). But to say that classical mechtech is modern is not to say that it is obsolete or no longer appears in our postmodern world. At the very least, one manifestation of the proliferation of aesthetics that is a hallmark of the postmodern is the fact that classical mechtech is now, if not *the* style, at least *a* style among others.

The hegemony that once amounted to nearly total domination that classical mechtech once exerted over western culture is broken. But mechtech is clearly present as one aesthetic among many, and as argued in this book, as one of three machine aesthetics still very much present in the cultural landscape. We now turn to the nine dimensions that will organize our discussion of the symbolic parameters of machine aesthetics, to understand the rhetorical potential for mechtech. Specifically, we will examine what is *meant* by the aesthetics of mechtech. We want to examine the symbolic resources within mechtech that comprise a repertoire for rhetorical application.

DIMENSIONALITY: DEPTH AND SURFACE

The dimensionality of mechtech involves an awareness of both surface and depth of the outer hull of the machine and its inner workings. The tension between the two is central to mechtech aesthetics. Sherry

Turkle describes this dimensionality as *transparency*: "Heaters, legos, erector sets, model trains. . . . The most significant thing they have in common is 'transparency.' An object is transparent if it lets the way it works be seen through its physical structure" (Romantic, 229). Perhaps the paradigm for mechtech dimensionality would be a 1960s style hotrod. These works of art displayed both a brightly painted (often in a shiny type of paint called "metallic"), smooth, and detailed sheet metal exterior and strategic glimpses of the machinery within (air scoops, lake pipes, side panels removed from the engine compartment). Mechtech dimensionality is *the machine in context*, gears and pistons within the frame of their housing, the dialectic between them being a part of the aesthetic. Edward Happold notes this bidimensionality of mechtech in saying that "for engineers, the aesthetic object . . . includes its internal working, *and* . . . the 'design' itself" (248).

Although machines have been part of human experience for centuries, their increasing complexity eventually led to the need to depict their internal workings visually, pictorially. This made it possible to study and to design machines such as clocks and locomotives. This development, which Roger Hahn traces to the seventeenth century, laid the groundwork for thinking about the dimensionality of mechtech: "The visual representation of machines forever stripped them of secret recesses and hidden forces. The learned reader now expected to see for himself how things worked" (146). Cutaway models and pictures today serve this purpose, of peeling away the shiny metallic surface of the machine in strategic places to reveal the depth of its inner workings.

An important part of mechtech's dimensional aesthetic is the revelation of how a machine works. Seeing past the skin and into its depth is a revelation, an epiphany, an avenue to knowledge. Yet the machine usually cannot work without that skin in place. As W. H. Mayall put it, "simple machines . . . can be appreciated because all the working elements can be seen and their functions understood" (11). This dimensionality informs the aesthetic of the tinkerer, who wants to probe the interior secrets of a machine to discover how it works. Writing of the dominant functionalist designers and architects of the 1930s, Van Doren asserted that

> if you probed into the past of most of today's artist-designers you would probably find them, in youth, mending discarded alarm clocks, building boat models, and tinkering up coaster wagons to run under their own power. They had the kind of curiosity that made them want to know what makes the wheels go round. (75)

As a path to knowledge, the dimensionality of mechtech is also a means to order. An aesthetic that stops at the humming steel or iron surface of a diesel engine does not seek the order that governs the spinning parts within. Knowing what is inside something is an act of ordering what is inside. Fernand Léger understood that this drive to order was part of the aesthetic shared by his fellow 1920s functionalists: "It is upon this sense of objectivity and precision in art, of an aesthetic based upon revealing the pattern, organization, and design beneath the chaos of experience, that both precisionism and objectivism are founded" (325).

Clearly, the nature of certain objects themselves hinders or facilitates a mechtech dimensionality. An object totally, inaccessibly enclosed in a steel or plastic case locates its dimensionality on its skin rather than on the skin-and-depth of mechtech. Yet the intentions and reading strategies of the audience also come into play, for the desire to go beyond the surface, to understand and therefore control the inside, partakes of mechtech bidimensional aesthetics even if the object in question seems to be high tech. Katie Hafner and John Markoff's study of computer outlaws and hackers notes that some hackers "latched onto computers as an extension of an adolescent compulsion to sit in their rooms and pull radios apart" (36). That motive appears to be a drive to treat the radio as mechtech. Turkle observes that many hackers also pick locks for fun (*Self*, 232), clearly an exercise in penetrating the impenetrable hard surface of a lock to understand and master the order within. These two examples of mechtech preoccupations among high tech computer hackers suggest that mechtech bidimensionality may be operating to some extent among those who know *so* much about a computer's inner workings that it has become a sort of 1990s hotrod for them, fingers skimming the surface keyboard so as to manipulate a thoroughly understood (for them) mechanical/electric interior.

SUBJECT

The second aesthetic dimension to be considered is that arising from identifying with the machine as a subject. What are the mechtech aesthetic possibilities from the machine in use, as a tool, as an extension of the self? We are considering the mechtech aesthetics of the factory worker, the handyperson, the home repair buff, the gun shooter: when that machine is a part of you, what aesthetic reactions are facilitated? A paradigm of this aesthetic dimension might then be the experiences of a factory worker at a machine, controlling a powerful tool in accomplishing a task.

When we as subjects merge with the mechtech machine, we enact, embody, and perform the aesthetics of the automated machine, of the robot. Nearly everybody has this experience to some degree, whether the machine we use is an electric mixer or a huge crane. Sheldon and Martha Cheney speak of this subjective feeling for the machine in observing that "the thrill felt in its harnessed power, the admiration for its marvelously precise functioning, its adaptability, its capacity for multiplying commodities, are part of a common experience" (24).

Several components of subjective identification with mechtech machines may be identified. The user of the mechtech machine gets into a groove. As a college student I had a job polishing grindstones in a factory. Like any job, it had the potential for soul deadening boredom. But when I was "into" the aesthetics of the experience, part of that appeal was the *rhythm* of the actions: insert grindstone from the left, lower the lever from the right, steady the interface between stone and polisher until the right amount of material is removed, lift the lever, remove the stone. Repeat, repeat, repeat. For Dewey, "esthetic rhythm is a matter of perception and therefore includes whatever is contributed by the self in the active process of perceiving" (163). In other words, it is a site of subjective identification with the machine, as both user and machine work in patterned tandem. Le Corbusier is correct, then, to praise "rhythm" in the aesthetics of machine work, which he sought to transfer to architecture and design (47–48). Mumford makes the obvious link between industrial rhythm and music (*Technics*, 202–3), a link that is clear in the "industrial" or "grunge" music that became popular towards the end of the 1980s, or in the earlier band Devo's ironically robotic tunes.

Siegfried Giedion notes that precise measurement and duplication of movement is key to mechanization of any industrial process, and that such precision and duplication was celebrated by the Italian Futurists in the early twentieth century (102–7). An important cognate of precision and rhythmic movement is *efficiency*. A mechtech tool used well is a tool used efficiently, every effort correlated precisely to a desired outcome. As Gelerntner put it, "The beauty of a proof or machine lies in a happy marriage of simplicity and power" (2). The time and motion studies that became popular during the first half of the twentieth century depended on this subjective aesthetic to create efficient work in all contexts. As Elizabeth Diller notes,

> scientific management, the practice of rationalizing and standardizing the motion of the working body conceived to expand industrial productivity, was introduced to domestic housework.

> Time-motion studies, developed to dissect every action of the factory laborer with the intention of designing ideal patterns of movement and, ultimately, the ideal laborer. (154)

The ideal industrial worker is efficient through *rationalizing* behavior; thus, *logic* is an important component of subjective mechtech aesthetics. Rationalized behavior is stripped of nonessentials. It is completely ordered to match the task. John Rajchman argues that Rudolf Carnap's logical positivism was well received by and consistent with the Bauhaus movement, for "no-nonsense logic seemed to fit with no-frills functional building" (165).

OBJECT

As a commodity or object to be gazed upon from a distance, the mechtech aesthetic object is geometric, oiled, metallic, and functional. Ann Ferebee summarizes the look and feel of the mechtech object:

> Machine-affirming artifacts tend toward geometric form and angular line, and are made by machine-based techniques utilizing glass, steel, concrete and other industrial materials. Hard and smooth-surfaced, they are often painted from a primary palate of red, yellow, or blue. (10)

Given the appropriation of mechtech aesthetics by many architects and designers during this century, this object dimension of the mechtech experience is quite widespread. Buildings especially, and to a lesser extent objects intentionally designed to be admired *as* designed, as crafted, rather than natural are apprehended from a distance, as objects. Since objects are perforce observed at some distance, the visual dimension or the "look" of mechtech is important. That look is one of "mathematical accuracy, physical economy, chemical purity, surgical cleanliness" (Mumford, *Technics*, 247). Ferebee notes that among cubist, futurist, and constructivist designers, "undecorated geometric objects symbolized the new machine age. Like machinery, they tended to be smooth-surfaced and fabricated out of steel, rubber and other industrial materials" (78). Lewis Mumford noted that for those early twentieth century designers, "face to face with these new machines and instruments, with their hard surfaces, their rigid volumes, their stark shapes, a fresh kind of perception and pleasure emerges" (*Technics*, 334).

A paradigm of the object dimension of mechtech aesthetics may be the experience of the person arrested by the display of new tools in the hardware store: row upon row of differently sized crescent wrenches, tidy packets of drill bits, or unscratched new vise grips. One does not yet think of using them as a subject, one simply admires their lines, their arrangement, their textures.

Admirers of mechtech as object often refer to its geometric quality: "delicacy, grace, and precise geometric beauty" (Brolin, 48). By "geometric" aesthetics is meant shapes following patterns of squares, rectangles, circles, and so forth—precise angles—with mathematically harmonious relationships among parts. Another way to express geometric aesthetic quality is to say that it is opposed to ornament of any sort: the object is stripped down to its functional essentials (Marcus, 47–66). In the absence of ornament, the precisely *measurable* dimensions of the essential machine are evident. Thus Le Corbusier sings of mechtech architecture that "steel girders and . . . reinforced concrete are pure manifestations of calculation, using the material of which they are composed in its entirety and absolutely exactly" (214).

Mechtech's geometric aesthetics are perhaps its strongest influence on the aesthetics of industrial culture generally, beyond the factory and tool shop. Fernand Léger observed this influence early in the century in noting that "more and more modern man lives in an order preponderantly geometric. All human industrial and mechanical creation is dependent upon geometrical laws" (45). Brent C. Brolin notes the influence of the machine on recent design: "Because of the machine's smoothness and geometrical form, as well as its essential 'functionality,' it served as the inspiration for modern forms" (46). He grounds architectural aesthetics in the geometry of the machine in arguing that "modern architecture's obvious sources of visual inspiration are the basically simple, 'clean' and coldly precise qualities that were first appreciated in the tools of modern times: airplanes, ships, heavy and light machinery and industrial buildings" (15). The beauty of the mechtech object is therefore akin to the beauty of mathematics (Florman, 135).

Surface texture is an important component of mechtech objects. The rapid, smooth motion for which many actual machines are designed requires a smooth, oiled look, as Harold Van Doren put it, "smart and gleaming contours" (17). Metal or a metallic look is often essential (Marcus, 94–114). We are referring not to the surface and depth of mechtech dimensionality, but to the look of mechtech objects, as objects of the gaze or touch.

Van Doren uses an interesting simile to describe the process of designing *any* object; it is to make "useful products more useful still, beauty clinging to them as you proceed almost by a process of accretion, like electrolytic deposits of metal in a plating tank" (55). Note the vehicle of the simile makes use of the smooth, metallic look achieved by plating. The frequently found reference to mechtech objects as "clean" is actually an aesthetic of surface texture. That clear, smooth texture is often thought to signify the simplicity and elegance of the mechanical process underneath as reflected in Le Corbusier's rhapsody over an automobile brake as "this precision, this cleanness" (121). Note that Mumford depicts mechtech aesthetic as "clean" due to its underlying functionality:

> The machine devalues age: for age is another token of rarity, and the machine, by placing its emphasis upon fitness and adaptation, prides itself on the brand-new rather than on the antique: instead of feeling comfortably authentic in the midst of rust, dust, cobwebs, shaky parts, it prides itself on the opposite qualities—slickness, smoothness, gloss, cleanness. (*Technics*, 353)

Elizabeth Diller notes that in the first half of the twentieth century, "the drive for efficiency often became an objective in itself. . . . The dust-and-germ breeding intricacies of the 19th century space were collapsed into pure surface—smooth, flat, nonporous and seamless" (154), a clean surface of mechtech metal.

AESTHETICS OF PRODUCTION

An important, complex dimension of mechtech aesthetics is its aesthetics of production. This is the aesthetic of the machine as it does its work—not the self-identified aesthetics of machine as subject nor the image of the machine as object, but rather the machine in productive motion. Efficiency, speed, uniformity, perfection, and specialization of purpose (or fragmentation) are components of this aesthetic mechtech dimension. Mumford notes that "expression through the machine implies the recognition of relatively new esthetic terms: precision, calculation, flawlessness, simplicity, economy" (*Technics*, 350), terms denoting the machine in production.

The term *perfection* is often used to describe what the mechtech machine does (Ewen, 194–95). Perfection in this sense means the ability to do precisely and exactly what the machine was designed to do, with no detectable variation. A drill press comes down in the same place, to the same depth, each and every time the lever is lowered; its productive activity is in that sense perfect. The affinity between machines and "measurement" (Le Corbusier, 67) allows this perfection, for a mechtech machine may be calibrated to move within a "perfect" range of tolerance.

The perfection of the producing machine is closely linked to its specialization of purpose, a situation sometimes referred to more pejoratively as *fragmentation*. Marshall McLuhan calls our attention to the fact that "mechanization is achieved by fragmentation of any process and by putting the fragmented parts in a series" (*Understanding*, 27). As Brolin notes, "a machine's sole purpose is to perform *a single* task" (48). Perfection is facilitated by that narrowness of purpose; a machine that is only calibrated to do one thing may be expected to do it perfectly.

We need to think about the ways in which mechtech aesthetic has influenced Western culture beyond the factory. The aesthetic dimension of the producing machine has been a cognitive and emotional template for wide ranges of culture. The mechtech aesthetic of fragmentation has a powerful effect upon the whole industrial process, and hence the whole society, within which mechtech reigns. Fragmentation of machine process has engendered the fragmentation of societies since the start of the Industrial Age. As Marshall McLuhan notes, "From the fifteenth century to the twentieth century, there is a steady progress of fragmentation of the stages of work that constitute 'mechanization' and 'specialism'" (*Medium*, 20). Ann Ferebee likewise notes that the Machine Age "broke the work process into separate fragments" (8). As Siegfried Giedion argues, "From the very first it was clear that mechanization involved a division of labor" (714).

Let us consider why the specialization and fragmentation of industrial processes fragments the worker. Of course, one obvious way in which that happens is that a worker is assigned to labor on but one fragment of the process along with her machine—they together make one isolated stop on the assembly line. The worker is thus separated from the whole process of production and from the finished product.

But McLuhan has another interesting explanation for why specialization and fragmentation also fragments the worker. It happens because a highly specialized machine uses only a part of the worker, isolating the value and effectiveness of hand or foot from the rest of the person. McLuhan notes that "what makes a mechanism is the separation and extension of separate parts of our body, as hand, arm, foot, in pen, hammer, wheel" (*Understanding*, 218). Hence, "human work and association was shaped by the technique of fragmentation that is the essence of machine technology" (McLuhan, *Understanding*, 23).

Industrial fragmentation in production thus creates fragmentation in the person using the machine, and beyond that in the culture at large. Ewen calls our attention to the fact that the goal of "scientific management" of life even beyond the factory is made possible by "the division of labor, the fragmentation of the work process, and the regulating function of continual measurement and observation" (189)—and note how well those terms apply to the nature of the producing machine. Yet many observers such as Ewen argue that this fragmentation reproduced itself across an entire society grounded in mechtech economic production, a process in which "fragmentation is aestheticized into the narcissism of mind and body" that is characteristic of industrial societies (192). This far-reaching claim bases a whole way of life in part on an aesthetic of fragmentation in the producing machine.

The producing machine also offers up an aesthetics of *simplicity*. Mumford notes that an essential part of machine aesthetics is "elimination of the non-essential" (*Technics*, 350). Machines "performed their specific tasks simply and efficiently. The aesthetic qualities of the machine—simplicity and geometry—became desirable in themselves" (Brolin, 33). Simplicity is sometimes expressed, as in Le Corbusier's work, as "economy"—of motion, effort, and energy (119). Mumford agrees in identifying machine aesthetic's "chief esthetic principle: The principle of economy" (*Technics*, 352).

Related to the idea of simplicity is the aesthetics of "uniformity, standardization, replaceability" (Mumford, *Technics*, 277) or "mechanical uniformity and repetitive order" (Mumford, *Art*, 43). "Standardization and interchangeability" of machines and their parts was central to industrialization and thus furthered uniformity (Giedion, 47–50). The machine's motions and processes are uniform; but of equal importance from the point of view of production is the fact that what it produces is uniform: the same bolt, the same spring, the same lever over and over again. This is the aesthetic of the production line, "an esthetic of units and series" (Mumford, *Technics*, 334): the appeal of identical copies flowing along (Ferebee, 34). Cheney and Cheney note that industrial technique means that "a million copies of one design may be produced with the necessity that the last shall be as perfect in efficiency and appearance as the first" (4). Ferebee offers as an early example of that uniformity the Colt revolver: "each part was reproduced with sufficient precision to make it interchangeable with the same part from another unit" (34).

When mechtech uniformity informs the aesthetics of a society at large, the result is often a valuing of "standardization" in architecture, labor, and industry (Gropius, 34). Gropius speaks in praise of the imposition of a "norm" to guide production, design, and life itself (34), one which would underlie the Bauhaus value of mass produced housing (38–39)—or as John Mellencamp sang, "little pink houses for you and me."

The producing machine is an image of *efficiency* and *regularity*, "predicated on the synchronicity of moving parts" (Ewen, 139). These themes are linked to the aesthetics of uniformity, perfect simplicity, and precision as well. A productive machine is efficient if regular in its motions. For that reason it is *rational*, the epitome of a planned and controlled process, the essence of *technique* (Ellul, Mumford), or as Baudrillard put it, "Technical connotation is epitomized by the notion of AUTOMATISM" (*System*, 109). And Alan Holgate reminds us that the image of efficiency and regularity is aesthetic:

> [W]hen objects are designed with logical purpose subjected to the discipline of economic efficiency, and design is pushed to the limit so that form is closely defined by natural "laws," we are able to recognize the resulting clarity and order of the form at a subconscious as well as an intellectual level. Such forms will appear beautiful to us. (222)

The beauty of control and regularity can easily become a model of social organization. Samuel Florman argues that engineers, perhaps most closely attuned to mechtech aesthetics, prefer such a world: "Every engineer has experienced the comfort that comes with total absorption in a mechanized environment. The world becomes reduced and manageable, controlled and unchaotic" (137). An aesthetic of control creates an interesting circular situation, in which machines instill in society an aesthetic of efficiency that then drives society to depend increasingly on more and more efficient machines. The rational machine perpetuates itself aesthetically, or as Mumford puts it, "The integral esthetic organization of the machine becomes, with the neotechnic economy, the final step in ensuring its efficiency" (*Technics*, 269).

In its efficiency and regularity, Mumford argues that the clock is the paradigm of the productive mechtech machine: "In its relationship to determined quantities of energy, to standardization, to automatic action, and finally to its own special product, accurate timing, the clock has been the foremost machine in modern technics" (*Technics*, 15).

And that image of the efficient, regular, controlled machine has surely informed the industrial societies built upon it; as Mumford also reminds us, "to become 'as regular as clockwork' was the bourgeois ideal" (*Technics*, 16).

The infusion of mechtech efficiency and regularity may be found, of course, in the schools of design and architecture most popular in the first half of the twentieth century. Ewen calls our attention to the "International Style" of architecture through the 1920s, which employed an aesthetic of "functional angularity" to project "economic efficiency and instrumental rationality" (143).

Mechtech aesthetics of regularity have informed social organization and practices in dimensions one might find surprising. Elizabeth Diller argues that the mechtech aesthetic of control and efficiency informed views of housework. Ironing, for instance, came to be seen as a process in which "the shirt is disciplined, at every stage, to conform to an unspoken social contract" following "the aesthetics of efficiency" (156). Sanford Kwinter agrees with the idea of the clock as the essential efficient machine, but notes the influence of that aesthetic on monastic life even before the industrial age: "the monastery is clearly nothing if not a prototype clock" that became a model for attempts to order and control all of society (193).

GENDER

It may seem strange to think of the aesthetics of a machine as implying a gendering of the machine. The notion becomes clearer when we understand that we are considering attributions made about both gender and machine under the rather specific cultural and historical conditions of patriarchy that have governed Western culture since before, but increasingly during, the industrial era. The aesthetic attributes of mechtech machines have allowed the machine to become an image that may express what patriarchy wants to say about *males*. On the other hand, males under a patriarchy that is also enmeshed in an economy dependent upon the machine will come to model themselves, or at least to aspire, after the aesthetics of the mechtech machine. If Samuel Florman complains that "man is weak, and yet the machine is incredibly strong and productive," it is only to identify the machine as a model to which *man* must aspire (130).

Aesthetic equations between males and mechtech are widespread, as in Mumford's reference to "the masculine reek of the forge" in steel mills as "perfume" (*Technics*, 209). That the perfume of the mill should be associated with the masculine may be due to the predominance of male workers there, but there are also aesthetic equivalences drawn so as to serve the purposes of patriarchy under industrial conditions. Ewen argues that early in the twentieth century "the machine, with its indefatigable 'arms of steel,' emerges as the prototype for virility, the mold from which the *new man* will be cast" (188). Prototypes for virility might conceivably be taken from any source; it clearly serves patriarchal interests to define the male as hard, "indefatigable," and indestructible in this way.

Cultural critics such as Ewen have noted that when masculinity is based on machine aesthetics, individual male bodies are expected to conform to those aesthetic standards: "The body ideal Raymond H__ covets is itself an aestheticized tribute to the broken-down work processes of the assembly line" (190). This is because, Ewen explains, the aesthetics of the ideal male body in the late twentieth century have come to mirror the fragmented, specialized nature of mechtech production: "The perfect body is one that ratifies the fragmentary process of its construction, one that mimics—in flesh—the illustrative qualities of a schematic drawing" (190). What this means is that, just as a given machine does one thing to perfection, and another machine does a separate thing precisely, so does the highly "ripped" image of the male body found in advertisements and in bodybuilding magazines tend to feature isolated muscles. Bodybuilders work to develop a given muscle or group to perfection before going on to strengthen another muscle, in an aesthetic homology with the separate action each machine performs on a developing object on the assembly line. The body then becomes a set of highly specialized muscle groups.

The resulting physique also shares mechtech's dimensional aesthetic, for the surface of the skin reveals the contours of each muscle below. For body builders, Ewen notes that "their bodies, often lightly oiled to accentuate definition, reveal their inner mechanisms like costly open-faced watches, where one can see the wheels and gears moving inside" (191), an arresting use of mechtech dimensional aesthetics to describe the male ideal. In this way, cultural expectations for these "most male" of male images have come to reflect a mechtech aesthetic. Thus, according to Ewen, "the 'masculine physique' has been the tablet on which modern conditions of work, and of work discipline, have been inscribed" (188).

The masculine aesthetic of the mechtech machine is also revealed in the feminization of its opposite aesthetic: *decoration*. The elaborate ironwork decorations, claw-foot stands, wrought flora and fauna found on earlier machines gave way to mechtech's aesthetic of simplicity and hard, smooth surfaces. The difference between those two aesthetics is often depicted in gendered terms. Writing during the height of mechtech's reign, Cheney and Cheney complain of the earlier aesthetic that "the machine was thus misused, masked, falsely frilled out with feminine and regal ornament" (48). Any sort of decoration on a machine is feminized and then dismissed by the Cheneys as "flowers and lace" (72). John Rajchman argues that in the 1960s the valuing of that gendered distinction changed, although the way in which it was gendered did not, for mechtech remained a masculine even if now a chastened patriarch: "The parsimonious distaste for unnecessary ornament, of which the positivists and functionalists had been proud, was said to be repressive and antifeminist" (166).

Another way in which mechtech aesthetic was masculinized was in its opposition to a feminized nature. An important tenet of the age of the machine has been the power that it gives specifically to subdue and control fecund nature. The relationship between eras that located power in nature or in the machine is described in gendered terms by Brian Setzer:

> The translation of earth-mother into force is immediately succeeded by yet another explanation of generation. . . . This new and miraculous body recovers not merely a male power of production but also projects the autonomy of that power. The "almighty machine" that displaces the colossal mother is a channel of energy. (31)

PERSONA: WARRIOR

To assign a persona to the aesthetics of a machine may seem fanciful. Yet I believe that different kinds of machines and the aesthetics they facilitate do strongly suggest certain kinds of characters or personae. I believe that some kinds of dramatis personae may reliably be found, with more or less predictable characteristics, in texts that are heavily informed by one kind of machine aesthetic or another.

African American folklore tells of John Henry in battle with a quintessentially mechtech steam engine. John Henry defeats the mechtech machine by being better at doing what it does; in other words, by being a better machine than the steam engine itself. It is not a coincidence that John Henry fought the steam engine, he did battle with it, for the persona suggested by mechtech aesthetics is, I believe, *a warrior*.

Every other aesthetic dimension of mechtech suggests the image of a warrior. Warriors stand as uniform copies row upon row, yet each soldier is capable of individual heroics in action. Warriors have the dimensionality of depth and surface; they are textured, equipped with armor, shield, bandoliers of cartridges, outer jackets and inner clothing. Warriors are trained to obey orders so as to do a single thing of the moment perfectly well, and continually if need be. The soldier who is running and shooting, running and shooting is similar to the laborer or the automaton attaching bolts to a chassis on an assembly line. As an object the warrior is smooth and gleaming with sweat or oil, hard and metallic against assault. Movement in the heat of battle is highly trained through efficient, repetitious drill designed to isolate needful movements for maximum efficiency. The warrior aesthetic is male, and its relationship to others one of violence. He revels in his physical power. The good soldier is one with his unit, which we shall see is consistent with mechtech's motivating context.

DOMINANT RELATIONSHIP: VIOLENCE

The mechtech machine is not gentle, nor does it work on material cooperatively. The drill tears violently into wood, the piston slams up and down at great speed, gears and cogs lock in tight combat as their wheels turn. A mechtech machine may be used to build a church, a hospital, a spiritual retreat house—but it does so with great violence!

At even the lowest, most "peaceful" level of use, the mechtech machine does violence in that it imposes control and mastery over an object, over nature, over people. The seemingly peaceful screwdriver grips the screw head and forces it to twist against the resisting wood. Mechtech has been a major instrument facilitating Western culture's desire to conquer and control nature and other cultures. Therefore, the violence of control may be seen even in Sanford Kwinter's view of the role of architecture as mastery: "Architecture's proper and primary function—at least in the modern era—is the instrumental application of mastery, not only to an external, nonhuman nature, but to a human—social, psychological—nature as well" (192). We learned earlier that the clock is an essentially mechtech machine in its precision and regularity. But Kwinter argues that it is also mechtech in the control (shall we say, violence?) it exerts over human affairs, calling it "a more or less generalized western technical apparatus of mastery—an apparatus whose power derives from its capacity to vanquish time by spatializing it" (194).

EROTIC: THE BODY

Erotic is the order of desire, and one of the dimensions of machine aesthetic. Mechtech erotic is the desire to exert power by extending the body through machines. It is an erotic of will located in the body, in love of motion, and the exertion of force. Mechtech erotic is the erotic of a *medium*, as described by Marshall McLuhan, which is an extension of the human body and its sensorium. One desires to move that boulder, and so uses a bulldozer. One wants to destroy that target over there, and so fires a gun. One wants to propel one's body faster, and pushes on the accelerator of the car. In each case, desire to extend the body powerfully is expressed through a machine.

McLuhan links a desire for physical extension to the mechtech machine explicitly in claiming that "during the mechanical ages we had extended our bodies in space" (*Understanding*, 19). Setzer makes a related claim when he argues that "the desire to go in a straight line couples the logics and the erotics of machine culture" (20). Note the linkage of physical movement, the forward *and unfettered* motion of the body following its desire, to mechtech machines. Herbert Marcuse argues that sexual desire (one dimension in an order of desire) in industrial societies is often channeled into, and expressed through, the machine (72–73). An everyday example of that truth lies in the acquisition of sexy automobiles as a way of expressing one's physical desire.

One recurring icon in mechtech culture is the robot, a machine that also facilitates electrotech aesthetics, as we will find later. Herbert Read makes the interesting observation that "the ideal of technology is complete automation—a machine that controls itself, without human intervention" (14). Is this not an image of a perfect erotic of the body, of the automaton's body? For if it acts without human intervention then it acts without external restriction or control as well. It is a body powerfully doing what it will. Clearly an implication of this observation is that mechtech erotic tends toward the not-human, the less-than-human, in its emphasis upon desire for brute physical power. This is, I think, what Sibyl Moholy-Nagy observes in the progressive depersonalization of Western culture, elimination of "the human element" by the robot dictatorship of remote-control systems, a boundless fascination with the interchangeable parts and sonic-kinetic combinations of electronic hardware (xvi).

MOTIVATING CONTEXT: THE FACTORY

The final aesthetic dimension of mechtech to be considered is its motivating context, the dramatic or narrative scene which would give a unifying, grounding cohesion to the whole. Mechtech aesthetics makes sense if one imagines the motivating context of a factory: each machine and each worker a unit in the whole. Every part is a necessary contribution to the smooth flow of the assembly line. Each part has its defined task to do, perfectly, and failure or deviation may harm the factory as a whole. Identity is taken from the plant as a whole: the individual machine makes sense in terms of what overall work is to be done. In a sense, the motivating context for mechtech aesthetics is The Borg. More familiarly, Mumford compared "machine civilization" to the regular, steady operation of the clock (*Technics*, 269).

The need for both human and machine to fit into the smooth operation of the factory is paramount. Mumford notes correctly that "the machine imposes the necessity for collective effort" (*Technics*, 280). Opposing two aesthetics, Florman notes the implications of mechtech: "A poet can be a rebel . . . but engineering and rebellion do not go together. . . . There can't be any engineering in a chaotic world" (180). One extreme application of mechtech's discouragement of rebellion was to be found in Stalinist society, where Martin Eidelberg tells us that "the power station and tractor symbolism was broadened to embrace the whole notion of ultramodern machinery, framing and energizing a society in which the individual was nothing and the state everything" (15). In that case, the state became a giant factory, thus defining the appropriate motivation for its citizen workers.

Life has conflict, of course, but Ewen argues that mechtech aesthetics have been used in actual factories to cover over the coercion and social pressures that suppress conflict in the name of production: "The look of the factory was aestheticized, divorced from any overt association with the coercive discipline or social conflict that were encompassed by factory life" (213). Mumford makes much the same observation in noting that machines produce a collective, anti-classist aesthetic, even as classism pervades industrial civilization (*Technics*, 353–54).

The equation between worker and machine in this motivating context is intentional, necessary, and historically grounded. Kwinter notes how the development of machines moved toward that equation: "The steam engine, rising upward through the world-system to the next level, combines with an economic flow reaching its own critical point (industrial capitalism) and is then combined with a cotton gin to produce a more complex entity: mechanized labor" (195).

RHETORICAL POTENTIAL

Our review of mechtech aesthetics has often suggested implicitly ways in which rhetorical uses might be made of its dimensions. Other scholars have noted the rhetorical applications of technology in discourse. For instance, Olson and Goodnight argue that today a new "technical sphere of discourse" is replacing nineteenth century "epochal" rhetoric which featured the machine. McCoy and McCoy argue that designers themselves are often at the forefront of those marshaling mechtech aesthetics for rhetorical purposes:

> What American designers are saying through their design reveals much about their vision of the relationship of technology to people. Some are rendering machines as serious and cerebral.... Others are attempting to humanize design by making it playful or casual.... Some are concerned with revealing the inner nature of the machine, the invisible processes within, with the intent of demystification. (17)

Here we turn more explicitly to consider some of the ways that mechtech aesthetics have been used rhetorically in the past to influence social order. We will consider ways in which the mechtech machine has been cast as a sign of progress, as democratic, as an underpinning of social harmony and efficiency, and as a moral paradigm. We will be noting especially ways in which several mechtech dimensions are joined integrally for rhetorical purposes. This section is thus an extension of the scheme of mechtech dimensions, and illustrates my assumption that to identify aesthetic dimensions is to identify a repertoire for potential rhetorical application.

Progress

The age of machines has been an age of progress. As humankind relied increasingly on machinery, expectations for economic and social progress grew. Machines and an ideology of progress were explicitly linked. As Brent Brolin explains, "Machines ... captivated the mind because, in an age that worshipped perpetual progress, they were the visible signs of progress" (48). This linkage was present from the start of the Industrial Age; as Roger Hahn explains, in the seventeenth century "machines were now directly linked to the expectation of progress through mechanical improvement" (146). Mumford is explicit in arguing that it was the aesthetic dimension of machines that gave people hope for developing "supreme skill and refinement" in life (*Technics*, 359).

The progress envisioned was of a specific sort, of course: material progress, control of nature, and economic prosperity. Ewen and Ewen note that in the early modern age, "the machine provided a 'reference-idea' by which the world could be understood, possessed, and mastered (2). This historic alliance between the machine and an ideology of progress occurs largely through the use of mechtech *aesthetics* as the feel and the look of progress. As Ewen notes, those aesthetics created a wide range of potential rhetorical application:

> The machine, its plain, angular lines and its seemingly perpetual forward motion, provided the most palpable metaphor for the force and progress of the machine age. As a "transcendental" symbol, it could be employed—simultaneously—by a variety of contending interests. (142)

American industrial prowess has allowed mechtech aesthetics a fertile rhetorical playing ground. Early in the twentieth century, Ewen argued that the machine gave rise in America to "a new aesthetic of power: calibrated, plainly geometric, unadorned, predicated on the synchronicity of moving parts" (139). Yeh notes that aesthetics of industry were used in the 1920s as a source of national pride and development: "Looking to the industrial world for inspiration for art and literature was a way of seeking indigenous roots for American civilization's emergence from provincialism" (3). Later, during the Great Depression, one use of mechtech aesthetics was to give the United States hope for progress and recovery despite the obvious failure of the economic system. Hugh Aldersey-Williams observes that "in machine-age America, design conveyed the optimism of the New Deal" (75)—note his focus on the aesthetic uses of mechtech. David E. Nye confirms that modernist aesthetics kept hope alive for eventual progress in the United States during the Depression (367). In sum, the look and feel of powerful engines, spinning turbines, gears and cranes, and thriving factories has been used to further an ideology of progress. That ideology has undergirded rhetorical appeals for nationalism, capitalism, and conformity to the demands of a sometimes unreliable economy.

Democracy

Mechtech machinery has facilitated a rhetorical celebration of democracy, often in contrast to the decorated excesses of Europe. Much of the linkage between machines and democracy has been aesthetic, as in the works of John A. Kouwenhoven. Paul Greenhalgh notes that modernist mechtech aesthetics was democratic in that its unadorned, undecorated "objects had to be self-consciously proud of what they were and how they had arrived in the world, much in the way that the democratised masses were encouraged to be proud of their origins and their status as workers" (9). Ewen notes that in the nineteenth century, "the unfettered simplicity that graced American tools suggested not only modernity, but *democracy*" (136). He goes on to stress the aesthetic nature of the link between mechtech simplicity and democratic access for everyone: "In each case, the tacit claim made by this technical aesthetic is democratic; it asserts that the prerogative of technical perfection is potentially available to everyone" (217). In contrast, Cheney and Cheney identify as "regal" any un-mechtech attempt at ornamentation (48).

Today, the power conferred upon anyone with the means to purchase a machine is democratic. The aesthetic that comes with such a purchase is an important dimension of that democracy, for aesthetics may be shared by all alike. McLuhan, in observing that "it is the *power* of the motorcar that levels all social differences" (197), points to the aesthetic of fast movement, the roar of an engine, the violence of a quick start. This common experience (horsepower being relatively cheap) is available to anyone with the ability to purchase even moderately priced cars. The same may then be said for the power of an electric rotary saw, a gasoline chain saw, or a snowblower; the *feel* of the mechtech engine is democractically and readily available.

Social Efficiency and Harmony

One of the more striking rhetorical applications of mechtech aesthetics, arising most clearly out of its motivating context of "the factory," is its linkage to demands for social efficiency and harmony. The linkage is often aesthetic; Ewen observes that "as the factory attempted to bring a calculable order to the realm of production, some reasoned that its motifs could inspire order and harmony in the volatile theater of everyday life" (138–39). Mumford notes that historically disorder and social contradictions are resolved symbolically by referencing the order and harmony of the machine (*Art*, 52).

Examples of mechtech utopians abound, visionaries who imagine social efficiency and harmony through better machines. Mechtech champions such as Herbert Read called for an "organic" unity between the machine and society and between the machine and nature in a utopian vision of how life ought to be (10–11). That unity was to be predicated upon making society, and a subdued nature, more like the machine. Cheney and Cheney use similar wording, arguing that industrial design can use aesthetics to reunify science and culture, "contributing to an organic culture such as the Western world has not known for several centuries" (6).

Perhaps the clearest rhetorical use of mechtech aesthetics is Le Corbusier's utopian vision of a machine-like society. Ewen notes that "his unvarnished mission—to coordinate disparate elements of production into a well-oiled, glitch-free apparatus—was generating devices and structures suited to the achievement of social harmony" (138). This harmony was envisioned in aesthetic terms as a well-oiled, smoothly running machine. Observe Le Corbusier's striking picture of the industrial utopia he imagined himself in during the early years of the twentieth century:

> Industry has brought us to the mass-produced article; machinery is at work in close collaboration with man; the right man for the right job is coldly selected; labourers, workmen, foremen, engineers, managers, administrators—each in his proper place. And the man who is made of the right stuff to be a manager will not long remain a workman; the highest places are open to all. Specialization ties man to his machine; an absolute precision is demanded of every worker. (254)

This vision is strikingly like Plato's *Republic* (consider Read's call for a machine-like society to follow "absolute or ideal Form," 11–12), and imagines a society headed by its own philosopher king: "The Engineer, inspired by the law of Economy and governed by mathematical calculation, puts us in accord with universal law. He achieves harmony" (7). Le Corbusier's view of social harmony through mechtech was contrasted with his complaint that in his day, "the machinery of Society [was] profoundly out of gear" (14).

Clearly, calls for social harmony and efficiency contain the seeds for rhetorical support of fascism and coercion. We are reminded that Mussolini made the mechtech trains run on time in Fascist Italy. The mechanical metaphor is pursued in John Rajchman's argument that Hitler's Nazi rallies were meant to run like clockwork in the hopes that the aesthetics of that mechanical order would affect social relations in Germany at large (194). Similarly, Florman notes that "a love of power, speed, and danger were the distinctive features of Futurism, followed by an impetuous glorification of war and fascism. Pride in the machine is obviously an emotion that has its dark side" (131). Against the grain of those who imagine a utopian mechtech society, Baudrillard argues that a social order founded upon the machine has no room for active, dominant humans; thus he makes a negative rhetorical use of mechtech aesthetics when he says, "Automatism amounts to a closing-off, to a sort of functional self-sufficiency which exiles man to the irresponsibility of a mere spectator. Contained within it is the dream of a dominated world, of a formally perfected technicity that serves an inert and dreamy humanity" (*System*, 110). An inert and dreamy public that expects machines to do its work may well be open to domination by government mechanisms.

Moral Mechtech

One of the most striking rhetorical applications of mechtech aesthetics has been the moralization of that aesthetics. Quite often we find mechtech aesthetics depicted as cleaner, morally better, more true, and so forth, clearly channeling mechtech toward certain rhetorical results. Note Mumford's conflation of ethics and aesthetics in arguing that "if we seek an authentic sample of a new esthetic or a higher ethic during the nineteenth century it is in technics and science that we will perhaps most easily find them" (*Technics*, 322). Gelerntner likewise mixes ethics and aesthetics together with machines in declaring that beauty "guides scientists toward truth and technologists toward stronger and more useful machines" (1). Brolin notes of early twentieth century modernists that "although they theorized about a *visual* art, they rationalized their visual choices in exclusively *moral* terms" (15). He grounds that morality in aesthetics when he argues that "the modernists' point of view that decoration serves no purpose and is therefore somehow immoral is totally in sympathy with Protestant ideals" (18). Just as those who would base social order on mechtech aesthetics argued for the *natural* superiority of mechtech simplicity and mathematical rigor, so that natural grounding became the basis for moral superiority as well. Brolin explains:

> Because of its rational origin, the machine was associated with the laws of nature that were knowable only through reason. This, in turn, led architects to assign moral qualities to the machine: it was considered "noble," "fundamental"—in the sense of reflecting "first principles"—and an "honest" expression of the times. (52)

A champion of mechtech such as an engineer may thus be imagined as, in Florman's words, "a leader in a great crusade" (4), conferring high moral status upon him or her.

Le Corbusier was one of the most persistent moralizers of mechtech aesthetics, as in his reference to "the mass-production house" as "healthy (and morally so too)" (13), or his insistence that "there is a moral sentiment in the feeling for mechanics" (119). In his view, because of mechtech influence on design and architecture, "the morality of industry has been transformed: big business is today a healthy and moral organism" (264). Note the moral tone in Le Corbusier's dictum, "Nor is it right that we should waste our energy, our health and our courage because of a bad tool; it must be thrown away and replaced" (17).

Likewise, Walter Gropius saw a "moral responsibility" in the work of the Bauhaus to disseminate mechtech aesthetics in society (89). In the 1930s, Cheney and Cheney claimed that "we have seen the beginnings of a new and truly creative art and the foundation of a new national and international culture which is a product of the mechanized world and rich in spiritual resources" (xi), clearly linking the moral weight of the spiritual to the aesthetic realm of art. Ewen finds similar moral application in another mechtech designer and architect: "The image of the machine, Laszlo Moholy-Nagy contended, reflected the spiritual force of the modern era" (141). Paul Greenhalgh explains that for Modernism in general, "Design was to be forged into a weapon with which to combat the alienation apparent in modern, urban society. It was therefore construed to be fundamentally a political activity, concerned with the achievement of a proper level of *social morality*" (8). As a result, Greenhalgh observes, "truth as a moral value was transposed into being simultaneously an aesthetic quality" (9).

CONCLUSION

Mechtech aesthetics have been the dominant machine aesthetic for most of the twentieth century. In the diverse, postmodern twenty-first century, it continues to be one important type of machine aesthetics among others. It should be clear from the sources cited in this chapter that most people during at least the first half of the twentieth century thought that mechtech aesthetics were a very *good* thing. Indeed, we might take that positive weighting to be one final aesthetic dimension of mechtech, since as we shall see, doubts and fears about all sorts of machines, mechtech or electrotech, are often expressed in the negative, ironic weightings of chaotech.

This chapter may well have illustrated the difficulties of keeping apart in tidy conceptual boxes the dimensions of mechtech aesthetic and the sorts of rhetorical uses to which they may be put. That conceptual fuzziness is central to the point I wish to make here, that the ways in which mechtech works as an aesthetic *are* the potential it has for rhetorical influence in different applications. Poetics and rhetoric, whether as practice or theory, have had a long history of intertwining. That is no less true when applied to machine aesthetics.

When we merged Dewey and Burke in the first chapter, we noted that we would be locating aesthetic experiences in both the perceiving audience and in the text that gave socially created parameters for acts of aesthetic perception. The usefulness of such a theoretical structure allowed us to think about a radio, although seemingly an electrotech object, in mechtech ways when regarded as a physical structure to be taken apart. In a similar way, someone who builds her own computer and is intimately familiar with how the shiny metal parts connect may be experiencing that computer through mechtech aesthetics. Therefore, as we move to our next chapter on electrotech aesthetics, we are not only considering different objects and actions but in some cases different ways to aesthetically experience the same objects and actions from which some people might derive mechtech, others electrotech, pleasure. Let us turn now to examine the significant aesthetic type of electrotech.

CHAPTER THREE

Electrotech: High Technology Machine Aesthetics

The term *high tech* is part of the indispensable vocabulary of our civilization. Every era has had its "high technology" machines; the wheel once held that pride of place! Much of the meaning of the term connotes the "newest" or "cutting edge" technology. But every age has had a "newest" technology that was replaced by something newer later on. In the late twentieth and early twenty-first centuries, Western culture also has its "newest" technologies and its "best" machines. Ask anybody on the street what "high tech" means and they will give you computers, cel phones, and so forth as examples. As Ellen Ullman put it, "We lead machine-centered lives: now everyone's life is full of automated tellers, portable phones, pagers, keyboards, mice" (146).

The "newest and best" technology today is the subject of this chapter, but not only because it is newest and best. Rather, today's high technology machines are qualitatively distinct from the high technology of the past. High tech today is not *simply* what is new or most valued in technology, it is qualitatively different from previous technology. For related reasons, today's high tech also grounds a particular range of aesthetic reactions.

I shall use "high technology" here to mean something more specific than just that which is new or best during the last hundred years. A high technology machine, in this study, is the kind of machine that is primarily dependent upon *electricity* to function. Hence the more descriptive term in the title of this chapter: *electrotech.*

Electricity has been harnessed for productive purposes for many years. Especially during the nineteenth century, industrial and domestic applications of electric power began to be widespread. Electric motors, lighting, and batteries came into use before the turn of the twentieth century. But many of those applications were incorporated as part of machines or industrial processes that were more likely to ground mechtech aesthetics. The internal combustion engine is experienced by many on mechtech terms, yet it incorporates electricity into its spark plugs, starter, and other components (even a diesel engine usually has an electric starter to turn the shaft over initially). The mere incorporation of electricity into its processes does not make a machine electrotech.

By electrotech machines, I mean more specifically machines that depend primarily, predominantly, upon electricity for their function. It is important to understand that these electrotech devices are machines, although they differ from mechtech. Samuel Florman reminds us that "computers and electronic devices . . . make less noise than lathes and power presses . . . but qualify as machines nevertheless" (127). Such machines began to appear late in the nineteenth century: the telegraph, telephone, phonograph, even the electric light may be thought of as an electrotech machine.

Now, around the turn of the twenty-first century, electricity is the main source of power, indeed, the soul of many machines that would popularly be called "high tech." Computers, fax machines, MRIs and a wide range of medical technology, and so forth—the list would be a long one—these are today's electrotech machines. They are what anyone would call "high tech," and they are what I think most people would regard as the newest, most cutting edge, and perhaps most valuable machines in our culture. Just as the mechtech railroad engine shaped so much of the Industrial Age, so electrotech machines have set the pattern for life today.

The electric nature of electrotech machines makes them qualitatively different from the "high technology" of earlier eras. Ann Ferebee reflects this qualitative break, and its aesthetic consequences, in her argument that "with the introduction of new energy sources, satellite communication and computer controlled information networks, we are now embarking on a second industrial age and the articulation of a new aesthetic" (5). Electricity is not simply the newest thing in machinery; it creates a different kind of machinery from what came before.

There are many distinctive characteristics of electric technology that we will consider presently, but I would like to sum them up under this one quality: electrotech translates the power and effectiveness of machines to the human scale.

A person using a mechtech jackhammer is controlling a power that she knows and feels to be beyond her own physical capacity because she is in direct contact with it. The teenager enjoying the roar and rumble of a mechtech souped-up V8 engine is at the controls of a monster with power beyond his own physical means. But to the extent that a machine is *electrified*, it brings enormous physical and cognitive power down to the human scale. Electricity allows machines to be quiet, personal, and small—even if they control powerful physical forces nearby or at another location. A person using a computer has harnessed vast calculating (cognitive?) power, yet it is controlled through the manipulation of the fingers. The technician at the control console of a nuclear power plant manipulates huge turbines *somewhere else*, but *here* all is calm, quiet, and under human control.

One compelling argument for understanding electrotech as being on a human scale is made by McLuhan, who sees machines as extensions of the human: "As contrasted with the mere tool, the machine is an extension or outering of a process" (141). Electrotechnology specifically is, McLuhan argues, an external recreation of the human central nervous system (53). Note McLuhan's equation of electrotech with the human scale: "Our new electric technology is organic and nonmechanical in tendency because it extends, not our eyes but our central nervous systems as a planetary vesture" (McLuhan, 136). Woolley agrees, in claiming that "television has become our eyes, the telephone our mouths and ears; our brains are the interchange for a nervous system that stretches across the whole world—we have breached the terminating barrier of the skin" (7).

Altheide notes that many electronic machines (telephones, computers, blenders, calculators) are now controlled by "keyboards," so much so that he calls the "keyboard" a medium in itself. In each case, the power to calculate, to communicate, to energize other machines and processes, is translated to the human scale of the fingers tapping on keys. The extent to which so many electrotech machines are operated by keyboards is a reflection of how electricity reduces the control of powers, processes, and other machines to human dimensions.

The human scale is key to understanding electrotech. Of course, many of the early electrotech machines were media of communication, signalling to us that the aesthetics of electrotech are informed by the aesthetics of communication in general. Communication is by definition on the human scale. McLuhan approached this idea of the human scale in electrotech from the point of view of communication media. McLuhan argued that a medium is any extension of the person. So television extends our eyes and ears, the telephone extends our ears and voice, and so forth. Yet in each case, the extension is so as to translate the power to send or receive messages vast distances to the human scale, to quiet movements of punching buttons or turning knobs. Norbert Wiener likewise notes the central connection between high technology today and communication: "If the seventeenth and early eighteenth centuries are the age of clocks, and the later eighteenth and the nineteenth centuries constitute the age of steam engines, the present time is the age of communication and control" (39).

On a related note, Ferebee argues that after World War II, industrial designers began to "dispose, conceal, miniaturize and otherwise dematerialize consumer products" (98). This remarkable statement points, I think, to the merger between electrotech machines and the human on the human scale. The computer, Ferebee implies, is so much an extension of the self, and so much on the human scale, that the "product" is beyond the metal and plastic boxes themselves. The product is the human on the one end and the information network on the other between which the computer mediates; the computer as product is thus relatively dematerialized, which can only happen because it is attuned to the human scale.

Since our concern here is aesthetic experience, I will take the ability of electrotech to translate and transfigure power to the human scale and thus to human experience and perception as key. In this chapter we will explore the aesthetic repertoire created by machines that allow the person to wield great powers of different sorts quietly, without physical strain, softly, within the range of accustomed human motion and behavior. Electrotech, unlike mechtech, is machine aesthetics without sweat.

Given the multifaceted nature of many machines today, we will regard electrotech aesthetics, as we did mechtech aesthetics, as a kind of experience facilitated by the potential within certain kinds of signs, rather than as an experience exclusively and singularly linked to particular kinds of machines. That is to say, a given machine might well have the potential to support a mechtech aesthetic at one time, or in one application, and an electrotech aesthetic at another. Clearly, some machines tend much more towards the potential for electrotech, as do some machines towards the potential for mechtech. But as we draw distinctions and identify what is peculiar to electrotech aesthetics, we will remain comfortable with overlap and simultaneity in terms of mechtech aesthetics. The sheer proliferation of machines of every sort today makes that kind of flexibility important if we are to understand how we experience these machines aesthetically. As Jean Baudrillard remarked, "Our urban civilization is witness to an ever-accelerating procession of generations of products, appliances, and gadgets" (*System*, 3).

It may be worth our while to remind ourselves that there *is* an electrotech aesthetic, noted by observers long before this present study. For instance, Richard A. Lanham summarizes at length his view of the aesthetic of the computer, which he sees as:

> Art defined as attention, beholder as well as object; thus an art that includes its beholder, and the beholder's beholder, an outward frame-expanding, an infinite *pro*gress rather than *re*gress; interactive text, that is, art and criticism mixed together, and so art and life as well; a continually shifting series of scale-changes, of what literary theory would call contextualism; a resolute use of self-consciousness to turn transparent attention to opaque contemplation . . . above all, a pervasive reversal of use and ornament, a turning of purpose to play and game, a continual effort not . . . to purify our motives, but to keep them in a roiling, rich mixture of play, game, and purpose. . . . Is this not the aesthetic of the personal computer? (50–51)

Note the extent to which Lanham's aesthetic depends on how the perceiver makes use of the aesthetic parameters offered by the machine. Furthermore, Lanham argues explicitly that this aesthetic undergirds "a new rhetoric of the arts" (14). Let us turn now to our scheme of aesthetic dimensions to understand the rhetorical potential inherent in electrotech aesthetics.

DIMENSIONALITY: THROUGH THE LOOKING GLASS—THE PERMEABLE MEMBRANE

Our discussion of mechtech dimensionality stressed the continuous, linked nature of depth and surface. The aesthetic experience is aware of both the iron contours of the engine and the pounding pistons within. Electrotech dimensionality is paradoxical: it makes surface and depth discontinuous, yet continues to span them both. It both creates and merges a duality of surface and depth. The aesthetic experience "stops" on the surface *or* the surface melts away and one enters the electronic depths within, yet each experience contains the other and depends on awareness of the possibilities of the other. A paradigm of electrotech aesthetics is the computer: one engages the external skin of the CPU, the monitor, the keyboard, and the speaker columns, with the expectation of the world one will enter upon using it—and then, one slips past that surface and enters in a fully involved way into the cyberworld "within" the machine, yet still using the surface tools and components of the machine (the keyboard, the mouse). Software engineer Ellen Ullman describes her engrossment in writing code in that way: "I have passed through a membrane where the real world and its uses no longer matter" (3).

David Gelerntner observes that cyberspace is a peculiar kind of space: "Software is stuff unlike any other. Cyberspace is unlike any physical space" (22). Yet he makes a compelling argument that a program or application in process is a machine fully as much as the hardware that contains it: "A running program is often referred to as a *virtual machine*—a machine that doesn't exist as a matter of actual physical reality" (24). The software engineer, Gelerntner claims, is not hampered by the notion of ordinary physical space, but often creates programs for nonphysical worlds, sometimes even for hardware that does not yet exist (24–25). Gelerntner's arguments support the idea of cyberspace, the world into which one enters using electrotech, as both a machine and a kind of strange space in and of itself.

Electrotech dimensionality is thus bidimensional. It is the experience of Alice, who both admired herself in the looking glass but then stepped into, through, and beyond that surface into a world discontinuous with her reflection. Yet she anticipated that world while looking at her reflection, and longed for her home world while in its depths. That looking glass experience is articulated by the hacker "Pengo," reported by Katie Hafner and John Markoff as saying of his adventures in cyberspace, "The chief thing for me was the adventure, suddenly being inside a movie" (245). It is interesting that Sidney Perkowitz summarizes the work of the scientist, that priest of electrotech, in precisely this way: "Anyone with scientific training has the power to look both at and through the surface of the world, the power to merge visceral reaction and cool analysis into heightened response" (63).

Let me briefly note here that the work of Michael Polanyi conceives of knowledge in a way that is quite congenial to electrotech bidimensionality. Polanyi imagines a paradigm example of an investigator probing a hidden cavity with a pick. Knowledge of the pick and of one's fingers, he notes, must become "tacit," or out of "focal" awareness, if one is to have focal knowledge of the cavity being explored. One *can* become conscious of the pick and fingers, but then one loses focal knowledge of the cavity. Similarly, to drive a car one must become focally aware of the road and the destination, *not* of the pedals, the wheel, and so forth. One can admire a book, the typeface used, the quality of the paper, and the fine binding; but to *read* one steps through that surface, that looking glass, into the world of meaning given by the physical book and the ink on its pages, going past the surface into the depths. Similarly, the novice before a computer might admire its sleek look. But there is a world to be entered by engaging that surface of keyboard, CPU, and monitor, a world that takes one beyond the surface even as one uses it. Another way to say this is that in electrotech, hardware tends to be surface and software tends to be depth. Speaking in Polanyi's terms, one can be focally aware of surface, or use the surface in ways that shift it to a tacit dimension that allows focal awareness of the depth within and beyond.

The Occluded Interior

One important factor of electrotech dimensionality is the way in which the surface of the machine "stops" the viewer, even if temporarily. As Turkle notes, "Computers (taken as physical objects) present a scintillating surface, and the 'behavior' they exhibit can be exciting and complex, but there is no simple, mechanical way to understand how they work" (Romantic, 229). Gelerntner notes the impossibility of understanding depths on the same terms that we understand the surface: "The world inside a computer . . . has no intrinsic appearance. Electronic information is stored in terms of voltage levels, which, strictly speaking, look like nothing" (63). The skin of electrotech machines often serves the important function of keeping dirt and water away from sensitive electronic components. But the skin, and the nature of the machine itself, also serves the purpose of making the interior of the machine a mystery. W. H. Mayall notes that "as machines became more complex, they acquired casings and housings which obscured the working elements" (15) so that "we cannot see the works nor, in many cases, the basic structure of many modern machines" (19). The mechanical guts of a computer or telephone are not something most people can readily understand (unless one is trained to do so, in which case one may well have a mechtech experience of those inner workings).

François Barre neatly summarizes the contrast in dimensionality between mechtech and electrotech:

> The watch with its wheels, the steam locomotive with its pistons, the crane with its boom and lifting gear, the bicycle . . . all are self-explanatory objects expressing with exemplary clarity a structure and its operation, a working aesthetic. The spread of electronics and computers, along with the increasing miniaturization of their components, is creating a universe of occultation in which things are no longer visible, no longer readily accessible to understanding. This occult design . . . generates abstractions such that nothing which moves or functions assumes discernable form. (11)

Similarly, Turkle compares mechtech and electrotech dimensionality in this way: "Legos, erector sets, model trains . . . The most significant thing they have in common is 'transparency.' An object is transparent if it lets the way it works be seen through its physical structure. The insides of a modern radio do not provide a window onto the patterns and interconnections of parts" (Romantic, 229).

The idea that electrotech machines do not allow easy understanding of their mechanical interiors is widespread. Deborah Lupton notes that "many people who use their PCs almost every day have very little knowledge of how they work, of what lies behind the bland, blue screen" (98). Marion Hancock argues that the trajectory of high technology means that "the technology modern products contain becomes more anonymous and less understandable to the ordinary person" (272). One interesting consequence of the occlusion of the interior depths of electrotech machines is that functionalism, that theme of many (but not all) mechtech designs, is no longer relevant to electrotech aesthetics. Mayall is speaking explicitly of high tech machines here: For computers, "we cannot apply that somewhat worn adage 'form follows function.' How could we show in a computer's casing the function it performs?" (24) What a machine does can no longer be naturally expressed in its form or appearance. Myriad tiny wires and microchips in one machine looks much like the same constellation of parts in another machine with a completely different function. Ralph Caplan notes the impact of this change on electrotech design aesthetics: "Functional Modernism emphasized the revelation of a product's working parts. But now there are likely to be no working parts visible to the naked eye" (10).

The occlusion of the inner workings of electrotech machines does not mean that they have no depth. We have been speaking of occluded hardware, of the mystery of the motherboard. Instead, depth becomes cognitive, psychological, experiential—*aesthetic*. One cannot go from appreciation of the physical surface of electrotech machines to appreciation of their physical inner workings, as one can with mechtech. The move to depth is a move to a different dimension, just as a move into the Looking Glass is a qualitative shift in one's involvement with the mirror. The depth of an electrotech machine is one's involvement, captivation, concentration, and engrossment with the task or application with which one is occupied. It is a depth revealed by engaging the electrotech machine and confronting its software, its logic, its function.

The software/functional interior of electrotech is also occluded, of course. Who can immediately use a new answering machine or cel phone before deciphering its mysteries? The surface and depth of electrotech is exemplified by the beginner's fascination, perhaps horror, with both the surface appearance and the mysterious logic of the machine. Ripping off the top of the CPU will not help the beginner understand its depths. Knowledge of the depths of the electrotech machine comes from the first steps and then from the growing engagement with the world within of word processing, web surfing, data processing, and so forth. Because that depth is of a different quality than the surface, electrotech dimensionality is one of boundaries and dualisms.

Boundaries and Dualisms

Because the surface and depth of electrotech dimensionality is discontinuous, electrotech supports an aesthetic based on boundaries and dualisms, even as it links and transcends them. Lanham makes an interesting distinction along these lines, although I think he is incorrect: He argues that people look *at* electronic texts and *through* printed texts, because the former is located exclusively on surfaces. Hence, "we look at the surface pattern, AT the design rather than THROUGH it" (43). One certainly may look at the design of a computer, or at the words and images on its monitor. But to become involved with the electrotech machine requires a cognitive, emotional, and aesthetic movement through that surface into the cyberworld of applications beyond it. This move from the electronic box before us into the world of its depths is described by Altheide in this way: "Media technology enables us to transcend our immediate spatial and temporal boundaries by gaining access to information located in a different time and place" (31).

Electrotech duality is found in observations going beyond, but I think informed by, its dimensionality. Turkle argues that "the computer is Janus-like. It has two faces" in that it is both a tool (an extension of its user) and a machine, which imposes its own logic upon the user (*Second*, 170). Elsewhere, she observes this duality: "Like dreams and beasts, the computer stands on the margins. It is mind and not yet mind; it is inanimate and yet interactive" (Romantic, 227).

The electrotech machine is the *whole* discontinuous experience between its surface form and the world of application contained within. In that sense, the electrotech machine is poised between boundaries and dualisms; it entails them yet overcomes them at the same time. It is in that spirit that McLuhan and Fiore observed that "our time is a time for crossing barriers, for erasing old categories" (10). Turkle thus calls computers "marginal objects" that lie between categories (*Second*, 31). Others note the paradoxical nature of an experience that crosses boundaries: it thus tends to overcome and merge those very boundaries. Mark Poster notes that increasing use of electronic media creates an "increasingly simulational" culture that blurs the boundaries between "originals and referentialities" (85). Mark C. Taylor makes the same point by saying that "the opposition between superstructure and reality becomes problematic" in a world of television, film, and cyberspace (104). Donna J. Haraway notes that electric technologies challenge the distinction between public and private life, merging that dualism in a depth experience that crosses both worlds as one surfs the net in the privacy of one's own home (177).

The notion of a simulation leads us to consider cyberspace, in which the distinction between image and referent, or representation and reality is confounded. George Slusser and Tom Shippey note that in cyberspace, "once the tyranny of mimesis vanishes, observer and observed are free to exist in the same space, which today is the space of *information exchange*" (2). We now turn to understand how the idea of cyberspace expresses the discontinuous but linked dimensionality of electrotech.

Cyberspace

The experience of engaging the electrotech machine as surface, yet using it to enter the depths within, is the experience of "cyberspace." Mike Featherstone and Roger Barrows offer this cogent definition: "Cyberspace is best considered as a generic term which refers to a cluster of different technologies . . . all of which have in common the ability to simulate environments in which humans can interact" (5). Cyberspace is the intersection between the controls at hand (the keyboard, the mouse—the surface) and the game, the web, the day runner, the spread sheet within the depths of the machine. Michael Heim argues that "the virtual environment sucks in its users with a power unlike any other medium" (*Design*, 68). Once in cyberspace "we interact with virtual entities, and we become an entity ourselves in the virtual environment" (70).

Notwithstanding, one does not truly become "lost" in cyberspace in the sense of losing a connection to the machine that is granting access to an application. Cyberspace is instead that balance between the video game controls in the hand and the dungeon of doom through which one is making one's way within the application. The polarity between the technology and the simulated environment, the world within, or the depth of electrotech is the nature of its dimensionality. Taylor very usefully puts the idea of cyberspace into perspective for us: "Cyberspace is already in our midst. In a certain sense, we enter cyberspace every time we pick up the phone, send a fax, or log on the net" (105). Precisely so; cyberspace is not being "lost in space" or trapped in a holodeck simulation, it is the tension between electronic machine and application.

Turkle expresses the tension between the operator of a computer game, sitting at the surface, sitting on this side of the mirror—and—the entering of the depths of a virtual reality. She observes that "video games are something you do, something you do to your head, a world that you enter, and, to a certain extent, they are something you 'become'" (*Second*, 66–67). But we enter that virtual depth while still spanning over into the surface of the machine itself, for as Turkle continues explaining, "It's more than thinking—in a way it is beyond thinking. The hand learns what to do and does it automatically" (68). On one hand, Turkle argues, one enters the virtual depths and becomes a character in a game (83, 85). On the other hand, the game becomes a [NB] "perfect mirror" of one's skill in manipulating the controls, the machine itself, the surface (88–90). Within cyberspace, "body has thereby been transformed into body image, a second-order electronic construct appropriate to the environment its consciousness inhabits" (Christie, 175). Yet anyone who has ever had a mouse fail while cruising cyberspace knows that the real body, on the surface of the machine, is inextricably linked to the cyberbody swimming in the electrotech depths.

Others, such as John Perry Barlow, note the bidimensionality of cyberspace in arguing that "to enter it, one forsakes both body and place and becomes a thing of words alone" (93). This observation is interesting for its parallel to language, which in many ways is also mere surface (marks on a page, sounds in the air) and also the means by which we enter into a cognitive and experiential depth beyond that surface. The words I am typing now on a computer screen's surface are also the means by which I enter into this book I am writing which is somehow inside the cyberspace of the computer.

Anthropomorphic Electrotech

We have noted that the physical interiors of electrotech machines are incomprehensible to most people, and thus the depths of electrotech dimensionality are not physical but cognitive, experiential, and aesthetic. The same may be said for our experience of other people: To understand human depths one probes not more deeply into guts and bone but into the qualitatively different dimension of thought, emotion, and value. Turkle has observed that the hidden physical interiors of electrotech machines encourages anthropomorphizing them in precisely that way: "the physical opacity of this machine encourages it to be talked about and thought about in psychological terms" (*Second*, 22, 272). Similarly, Deborah Lupton notes that people attribute emotions to computers and in turn react emotionally towards them (104–5). Although he is not talking specifically about electrotech machines, Ewen's observation about streamlining is germane here. He argues that the advent of streamlining in the 1930s and 1940s gave machines a "soul" precisely because it occluded their interiors and covered them up—in much the same way that electrotech surfaces cover up their altogether incomprehensible interiors (145). We shall see later that anthropomorphism grounded in electrotech dimensionality facilitates the rhetorical potential of the *cyborg*.

A MALE SUBJECT: LIMITS AND EXTENSIONS OF THE WILL

In our discussion of electrotech gender, I will claim that it is androgynous. In preparation for that claim it is important to understand the ways in which electrotech is gendered differently as subject and as object or commodity. As subject, electrotech is male; as object, it is female. That difference has a lot to do with its androgyny.

As with mechtech, any claim about the engendering of a machine aesthetic must be understood within the context of a patriarchal culture. Patriarchal assumptions locate will and mastery in male subjects more than in female; males are trained to desire mastery and to cultivate the assertiveness and aggression needed to achieve it. Michael Heim observes this bias in electrotech towards the patriarchal values of control and manipulation: "[Virtual reality] has a tilt toward manipulation, even a latent tendency toward aggressive, first-person attitudes" (*Essence*, 29). One may see the truth of that in the persistently violent, quest-and-conquest orientation of most computer games and simulations on the market. Electrotech machines both require and exercise the will; they are fundamentally, as Turkle reminds us, instruments of *control* (*Second*, 210). Electrotech's triumph of the [male] will is seen in Mark Setzer's observation of "the violent immediacy promised by communication and control technologies operated by the electric signal or button" (11).

Because they are discontinuous, the surface and depth of electrotech dimensionality are not easily crossed, nor are they apparent on the same level of experience and understanding. When one is using an electrotech machine, joining with it as a subject, one must will to push beyond the surface into the always initially difficult depths of cyberspace within. Once there, it is will that is exercised as one cruises from one application to another, one cyber reality to the next. It is in this sense that Slusser and Shippey describe electric media as "extensions of experience" (1). This will to enter cyberspace, the desire to exercise the will while there, is what Turkle means in saying that "the computer . . . is a powerful projective medium" (*Second*, 14). The will to push on into the wilderness of cyberspace is the motive of the explorer, as noted in Hafner and Markoff's description of this hacker: "Lenny had always enjoyed the aspect of traveling through computer systems that made him feel like a fearless explorer. He liked the idea of having computers throughout the world at his fingertips" (110).

We should note that technologies encourage the development of subjectivities in their own images. As Taylor puts it, "As technologies develop, subjectivity shifts. In the culture of the simulacrum, or simcult, the thread from which structures of subjectivity and fabrics of experience increasingly are woven are fiber optics" (104). Barbara Kantrowitz argues that men and women see computers differently. Women see them as tools to be used, men as extensions of their own power (135, 140). Her argument supports the notion that men's experience of computers is *more likely* to be subjective, more likely to be a bond between machine and user. If the subjective experience of electrotech exercises male subjectivities through extension of the will, electrotech aesthetics may become an instrument in support of traditional patriarchy, a means of shoring up hegemonic male subjects. In an earlier age,

David E. Nye notes how the development of electrical technology became a metaphor for power that was used to describe individuals, particularly leaders: "Electrical terminology suffused popular language with metaphors of power, and in the Depression years Americans expected their leaders to be 'live wires' who would recharge the economy" (340). Since nearly all of those leaders were male, this serves as an example of electrotech's reinscription of power in male subjects.

One way in which subjective identification with electrotech exercises the will is that electrotech machines tend to be instruments used by and aligned with the individual. Although that is true of many mechtech machines as well, electrotech machines are almost entirely for individual use, and therefore for the furthering of what the individual wills. As Lanham puts it, "the small computer reflects and enables our individual, idiosyncratic needs and purposes as no device has done before" (199). Success at electrotech then becomes an individual success and triumph; one computes because one *can*, as a way to show one can do what one *will*, as much as for achievement of independently established goals. This is what Turkle means in observing that for hackers, "their pleasure is in manipulating and mastering their chosen object, in proving themselves with it" (*Second*, 201). The theme of individual, personal mastery is also found in Ewen's observation about electrotech design:

> The design of many products—particularly appliances and other electronic items—suggests that with the purchase of the product, *you will have your hands on the controls.* In a world where a genuine sense of mastery is elusive, and feelings of impotency abound, the well-designed product can provide a symbolism of autonomous proficiency and power. (215)

Electrotech subjectivity exercises the will through the newest-of-the-new technologies that require activity and assertiveness. Featherstone and Burrows note this interesting contrast: "The privatized retreat into television and video—essentially passive, non-interactive mediums—has been followed by engagements with increasingly interactive technologies: camcorders, multi-media interactive CDs, computer games and so on" (13). Mark Poster seconds this remark by noting that "subject constitution in the second media age occurs through the mechanism of interactivity" (88). Let us remember that earlier we noted the special role of electrotech in creating subjects; Poster's argument then supports the idea that traditionally active, assertive male subjects may be encouraged by electrotech.

A FEMALE OBJECT: FEAR, MYSTERY, AND CURVES

Just as the subjective experience of using an electrotech machine is shaped by patriarchal assumptions about being male, so is the experience of regarding electrotechnology as object or commodity shaped by patriarchal assumptions about being female. Electrotech objects are engendered as female under patriarchy, and thus acquire meanings usually attributed to women under patriarchy.

For biological and social reasons, female bodies are more rounded and curved than are male bodies—at least in popular myth and imagery. The surfaces of electrotech objects tend to be more curved and flowing than is the case for mechtech objects. Marion Hancock cites this as a general trend in an increasingly electrotech culture: "As metal bashing has given way to electronics, industrial design has moved . . . from the straight-sided to the curved" (272). The computer at which I sit as I write this has rounded corners on the CPU, monitor, keyboard, and mouse. There is not a sharp angle anywhere: even the stand on which the monitor sits and the pad underneath the mouse have rounded corners and gentle curves. So does the cordless telephone that sits nearby, and my pocket organizer that is lying in my briefcase.

The smoothness of electrotech objects has come to inform other signs in the culture, as the meanings and connotations of high technology are appropriated for rhetorical purposes. Ewen notes that "the unbroken smoothness that characterizes much corporate design bears out the idea of technical perfection, of an aesthetics beyond the capacities of human creation" that is now signified by much electrotech (213). Consumer products that are not especially high tech may also be designed with smooth and flowing shapes, as in Thomas Hines's observation that the rounded shapes of satellites came to influence the shapes of appliances such as vacuum cleaners in the late 1950s and 1960s (131–32).

The smoothness of electrotech shapes is homologous with the smoothness of their surface textures. Electrotech is relatively more likely to be covered with a matte or low gloss plastic skin; the bright shine of stainless steel is more often found on mechtech devices. A recurring idiom in which the smoothness of all machine aesthetic surfaces is expressed is that of hygiene; and so Frances Bonner describes computers as typically depicted (in media representations) in "very clean settings" (195).

The popular patriarchal consciousness has traditionally depicted women as *mysteries*. The themes of women-as-different, as unknowable, as changeable and thus unpredictable, as directed inward, all run through the popular media representations and social knowledge that are a part of a patriarchal culture. Woman-as-mystery may well be a myth perpetuated under patriarchy because of the relative hiddenness and occlusion of the primary female sexual organs. The parallel idea of the electrotech machines, especially the computer, as "complex and unpredictable" in anthropomorphic ways is, of course, widespread (Turkle, *Second*, 194).

If woman-as-mystery is a patriarchal myth, then other mysteries may come to be equated with, or expressed as, female. There are surely few more daunting mysteries to much of the population than electrotechnology, especially computers. Sadie Plant explicitly argues that women, cyberspace, and software are metaphors for one another, saying that "misogyny and technophobia are equally displays of man's fear of the matrix" (62). Turkle likewise sounds the theme that people "fear the machine as powerful and threatening" (*Second*, 13). Such fear is likely to be of the machine as object rather than as subject.

Streamlining

When considering the rounded, female aesthetics of electrotech, it is important to consider the place of streamlining in the design of machines. Streamlining is, of course, the construction of machines and of casings or covers for machines so as to offer little resistance to an air stream. Smooth and rounded shapes are therefore essential components of that aesthetic. Streamlining became increasingly popular during the 1930s and afterwards; Sheldon Cheney and Martha Cheney mark its rising popularity in their 1936 work, for instance (17–18). Some have argued, as does Ewen, that streamlining a machine is a "lie," a retreat from machine aesthetics towards forms that are "seamless and rounded, organic" (145), and a denial of function: "Refrigerators, toasters, radios, water heaters, and pencil sharpeners were all imprinted with a modern look. The aerodynamic capacities of each were irrelevant to their use, but relevant to their image of being up to date, relevant to their sales" (148).

I disagree with Ewen; I believe that streamlining is a transitional aesthetic between an era dominated by mechtech and one dominated by electrotech. The aesthetic foundations of streamlining appeared long before its functional uses in missiles; Paul Greenhalgh notes that Modernism earlier in the century gave rise to the International Style with its "sleek simplicity" and "white plastic" components (4). Martin Eidelberg also notes the way in which streamlining began in but moved beyond Modernism: "Streamlining was the branch of Modernism that combined the principles of aerodynamic engineering with the functional geometry of the International Style" (72). This aesthetic of simplicity and geometric form contributed, as we noted in chapter two, to mechtech's abhorrence of decoration. In a sense, then, streamlining as a transitional aesthetic borrowed the simplicity and geometry of mechtech design to point towards increasing electrotech functionality.

Streamlining became dominant as airplanes and sleek, diesel locomotives were becoming a significant "high tech" standard, soon to be followed by missiles and nuclear submarines. Cheney and Cheney were explicit, writing in the 1930s that the airplane was "the most conspicuous symbol and inspiration of the age" for machine design, and the model for streamlining (102). For those planes, streamlining is entirely functional. Electrotechnologies also assume increasing importance in airplanes; they were harbingers of our own highly electrotech era. It is true that streamlined design was applied to nonelectrotech machines such as manual pencil sharpeners, as Ewen points out. But that was for rhetorical purposes of borrowing the positive connotations of a transitional style, an aesthetic that was the "highest-tech" of its day, pointing towards electrotech.

Streamlining was thus transitional especially in its rhetorical applications. Streamlining, as Jeffrey L. Meikle points out, stood for the future (188). It captured the romance of jets and rockets, as Hine notes (88). David E. Nye observes that "in the 1930s streamlining became especially popular; it borrowed the terminology and the shapes of aerodynamics to glamorize automobiles, refrigerators, mixers, dishwashers, and other appliances" (354). Evert Endt and Sabine Grandadam note that streamlining "quickly became dominant, for it expressed the speed and optimism of the machine economy" of the 1930s, which was in transition to electrotech (33–34). Jocelyn de Noblet describes this intentional strategy of design in the 1920s and 1930s: "To seduce consumers, the publicists of Madison Avenue and a few designers thought it indispensable to develop an explicit brand of modernity. Aerodynamics appealed to them as the visible symbol of speed and before long a formalist mode of aerodynamics had been christened as the streamline style" (25). Meikle agrees that streamlining connoted "a vision of a smooth, frictionless, machine age future" (182). In the 1950s, Meikle notes, "objects as varied as cameras, radios, hamburger stands, and even petrol pumps took on sharply flared silhouettes, ultimately in imitation of the jet fighters of the era" (191). Writing during that era, Cheney and Cheney note approvingly that "everywhere, through the air, on rails, by land and water, there is the established point-counterpoint rhythm of smooth, gliding, mechanized travel, making its appeal to the senses as power dynamized, dramatized" (97).

Streamlining is interesting in its androgyny, as befits a transitional style. It includes both the long, phallic designs of airplanes and bullet trains as well as the rounded female shapes of 1930s toasters and 1950s vacuum cleaners. Today's female shaped electrotech objects have retained the curves but by and large lost the phallic component of streamlining.

AESTHETICS OF PRODUCTION

This aesthetic category is the experience of the electrotech machine at work. What is key to this aesthetic is the beauty of a *mystery revealed*. There is, as Deborah Lupton has observed, a "mystique" about the computer; people "have very little knowledge of how they work, of what lies behind the bland, blue screen" (98). The occlusion and strangeness of the electrotech machine must be mastered, opened up, and revealed at least in part for it to be in production. Thus, the aesthetic of the productive electrotech machine is that of a mathematical proof, an equation solved, a code broken.

Sherry Turkle's work with those who use computers the most for production has revealed this experience of a deep mystery revealed. She found that "they spoke about 'cognitive play' and 'puzzle solving,' about the 'beauty of understanding a system at many levels of complexity' " (*Second*, 167–68). Hackers, professionals, and the true aficionados love the computer's "potential for creating worlds of transparency and intelligibility" out of mystery (171). She found that these users have "the impulse to find a way to a sense of intimate understanding of the machine" (187). These observations point to an aesthetic that must occur in use, as the computer does what it is meant to do: creation of transparency and intimate knowledge from what was once confusion and frustration.

Much of the aesthetic of revealing the mystery of electrotech has to do with understanding the structure of ideas and processes that constitute cyberspace. Computer workers discover not just mysteries but intricately structured mysteries. As Turkle observed, "one can discover truth or something like that by looking through the patterns that connect this device" (*Second*, 254). She notes the ways in which "the hacker's computational aesthetic with its emphasis on intricacy of structure carries over to musical taste" (219) such as a love of Bach. The mystery that is revealed thus creates not only mastery but "mastery over complexity" (225).

The result of a mystery revealed is speed, power, and control; here the aesthetic of electrotech production is linked to the male aesthetic of electrotech subjectivity. Hackers are, according to Turkle, "the 'kind of people' who demand perfection and are compelled by the controllable" (*Second*, 200). The ability to control the electrotech machine while in production is facilitated by the ubiquity of push-button controls (Hine, 123–25). The push button "told its user that the machine in question was competent and complex" in production (Hine, 124). Morris Asimov finds the "rapid pace of technological development" today is mirrored in what we expect from electrotech production: "bolder, faster improvements" (2). Turkle notes the appeal of control and perfection in video games (*Second*, 88–89); a control and perfection achieved, as any player knows, once the mystery of the game is solved. Stuart Hampshire celebrates the computer's "copious and definite memories—efficient, clear, and unconfused" and compares them to the aesthetic experiences of ballet (254); his language aestheticizes precision and clarity, the results of revealed mysteries. Similarly, W. H. Mayall calls for computer designs that will reveal their "preciseness" (24). We should note that if the electrotech object is feminized in embodying mysteries, then the aesthetics of electrotech production seem to reproduce patriarchy in using subjective male will and control to penetrate those mysteries.

ANDROGYNOUS GENDER

The androgynous engendering of electrotech is apparent along several dimensions. The clearest sense in which electrotech is androgynous is the way in which it skews male as subject yet female as object, as we have seen earlier in this chapter. We have noted before how the categories of machine aesthetic used here are not totally distinct and separate. The concepts of machine as subject and machine as object are useful yet difficult to keep completely apart; we are likely to shift back and forth from a subjective to an objective experience with any sort of machine during a particular experience of it. In this sense the male subjectivity and female objectivity of electrotech creates a real androgyny rather than simply ambivalence.

In discussing electrotech dimensionality earlier, we noted that electrotech implies a crossing of boundaries between seemingly discontinous and incompatible dimensions. In that sense, electrotech is formally androgynous, in that androgyny is a crossing and transcendence of boundaries ordinarily taken to be discontinous and incommensurable. Lupton notes that going into cyberspace is in a sense leaving the body: "A central utopian discourse around computer technology is the potential offered by computers for humans to escape the body" (100). Yet since human bodies are engendered, socially and physically, the inner cyberspace dimension of an electrotech is thus a kind of androgyny.

PERSONA: THE MAGICIAN

The expression "computer wizard" is one of many slang indications of the mysterious power immanent in the master of the electrotech machine. It is part of our shared social knowledge that a special and arcane wisdom is required to engage electrotech. That knowledge is within the grasp of everyone at a rather low level: we can all do parlor tricks and cast tiny spells with our pocket organizer and cel phones. But "hackers," the real masters of the (Black?) Art, have grasped a level of knowledge beyond our reach. We also noted earlier how the electrotech machine in production facilitates an aesthetic of revealing mysteries and secrets, as one finally masters this or that application. These aspects of electrotech aesthetics make the magician the dominant persona linked to this kind of machine. Hafner and Markoff put it explicitly: "Hackers have become the new magicians," and they quote Arthur C. Clarke as saying, " 'Any sufficiently advanced technology is indistinguishable from magic' " (11).

Turkle observes precisely this persona in the aesthetic experience of one computer expert: "His was a magician's fantasy because what he was looking for ideally was doing something small, like touching one key or typing one character, and having the whole system come alive" (*Second*, 176–77). She describes another hacker in terms that would fit the mage precisely: "He had learned a new language, felt himself to be the holder of esoteric knowledge, and several nights a week were taken up with sharing its collective rituals" (174). This observation is consistent with our focus on electrotech as linking vast powers to the human scale, so that a wave of the hand or of the wand (keyboard?) it holds puts great events and consequences in motion. Kevin Robins similarly describes computer hacking as "the aesthetic of fantasy-gaming" (139), in which magic looms large.

DOMINANT RELATIONSHIP: SEDUCTION

The small, human scale on which electrotech works suggests a kind of relationship based not on the violent overpowering of mechtech but on seduction. Successful engagement with the machine requires coaxing it, working with it, understanding it, coming to an insight about the machine on its "own terms." Hafner and Markoff point to the interesting use of the term "social engineering" in hacker jargon to describe the practice of duping, persuading, and charming people (60, 123). Note the equation between rhetorical seduction and the engineering upon which electrotech depends in that usage: if persuasion is a kind of engineering, then engineering is a kind of persuasion, and thus a seduction. Like the magician who must understand the secrets of the universe so as to manipulate them, the electrotech user probes the machine's mysteries not to overwhelm and overpower them but so as to seduce them into alignment with one's own purposes. Ullman develops an extended metaphor along these lines in which learning a new computer or operating system is seen as a courtship (101–2). As with any seduction, the electrotech experience always leaves more to be uncovered, more to be learned, more to be explored.

Of course, seduction is also strongly suggested by a relationship going "the other direction": electrotech seduces its user. The innocent who began by tentatively surfing the web half an hour here and there ends by being wholly the machine's, lost in cyberspace for hours at a time. As Heim puts it, "no artifact so insinuates itself into the inner sanctum of the mind as computer-generated images" (*Design*, 70). Millions of people cannot now get through the day without watching television, nay, television operated by remote control; what is that if not an ongoing seduction by the electrotech machine. People who did not know that they needed cel phones now cannot live without one.

AN EROTIC OF THE MIND

We observed in the second chapter that mechtech aesthetics embraces an erotic of the body, extending and satisfying desires, the will, through physical action. Electrotech depends much less on gross motor movement, bringing its power to satisfy desire down to the human scale. As David Altheide describes it, "Keyboards have expanded rationality by tying more activities to a communications format" (212), and keyboards are of course a hallmark of electrotech. Electrotech embodies an order of desire grounded in the mind; it is thought that does the work, that generates the desires, that are then achieved by the electrotech machine. L. Casey Larijani describes electrotech as an order of desire in the mind: "The technology allows you to create an environment and participate in an experiential script of your choosing. You could become immersed in any way you wish" (x)—as long as you desire within the order of the mind. Robins reminds us that electrotech *does* exercise an erotic: "Artificial reality is designed and ordered in conformity with the dictates of pleasure and desire" (144).

Those who see the world in terms of a split between mind and body locate electrotech squarely in the former. Turkle quotes Burt, an MIT student, in exactly this way: "I think of the world as divided between flesh things and machine things. The flesh things have feelings, need you to know how to love them, to take risks, to let yourself go. You never know what to expect of them. . . . Math, you could get it perfect. Chemistry, you could get exactly the right values when you did your experiments" (*Second*, 198). Turkle finds this attitude typical: "The prototypical hacker's taste . . . tends not toward a sensual caress but toward an intellectual contact" (*Second*, 219).

John Christie argues that an important electrotech aesthetic is rejection of the body, of the physical, in favor of the mind:

> The cyberspatial body, all nerve and cerebral cortex, persists as a subjective location within the matrix, but it moves with the mathematical precision and electronic velocity demanded by survival within the matrix and is shorn of the demands and conditions of meat: hunger and age do not matter. (174–75)

An order of desire located in the mind is preferred among electrotech devotees to one located in the body, for as Larijani notes, in cyberspace "we can zoom or loom upon command and examine virtual things from angles unheard, through infinitesimally keen computer-aided eyes" (71). Likewise, Joseph Tabbi notes that "in [the science fiction classic] *Neuromancer*'s adolescent mysticism, immediate experience through the body is mostly a 'meat thing' to be rejected in favor of the cerebral pleasures and the 'consensual hallucination' of cyberspace" (212) which is, of course, a site in which erotics are expressed through the mind. Jocelyn de Noblet expresses this split as one between "volumes" and "platitudes," or mechtech and electrotech respectively: "Electronic components . . . are the reason for the gradual disappearance of the functional aesthetic based on volumes. This new phenomenon takes the form of an aesthetic of platitutes" (249), or abstraction and simulations within the mind.

Many observers have noted, perhaps *ad hominem*, a disregard among hackers for their own physicality, in favor of an erotic of mind (e.g., Nigel Clark, 118). Hafner and Markoff note hackers' physical awkwardness and nerdiness (17, 26), often coupled with poor diets and hygienic practices (29). Lupton agrees, in noting that in contrast to their exercised minds, hacker bodies "are soft, not hard, from too much physical inactivity and junk food" (102). Similarly, Ullman describes the bodies of herself and her fellow software engineers in quite unflattering terms, yet rejoices that "our bodies were abandoned long ago, reduced to hunger and sleeplessness and the ravages of sitting for hours at a keyboard and a mouse" (4). Turkle also notes hackers' perceptions of themselves as ugly, which she argues is encouraged by the fact that "our society accepts and defensively asserts the need for a severed connection between science and sensuality" (*Second*, 197). Such a dichotomy must then locate erotics in science, in the mind.

MOTIVATING CONTEXT: THE WILD WEST

We have seen that the context implied by a machine aesthetic is an important part of the total aesthetic experience, for it shapes motives and expectations. I argue that "The Wild West" is the motivating context for electrotech. By that I mean to call attention to the popular myths of the Wild West, of the frontier. Those myths identified a vast, decentralized social structure. Pockets of "civilization," of firm control certainly existed in the forms of towns. Communities arose here and there, whether grounded in village, ranch or farm, or outlaw gang. But the dominant image is one of freedom with loose connections: the lone horseman, bound by the "code of the West," riding here and there at will on the great frontier. Lewis Shimer makes the metaphor explicit: "The console cowboy . . . is an adolescent male fantasy to ride unfettered on the consensual range of the matrix, to shoot it out with the bad guy, and finally to head his chrome horse off into a sunset the color of a dead television channel" (23). Cyberspace, whether within the television, the computer, or the fax machine, is not quite, as Larijani puts it, "a virtually lawless world" (71) but it *is* decentralized and loosely structured. Comparing cyberspace to libraries, Ullman similarly notes that "the current reigning ideology of the Internet is strictly opposed to the idea of a librarian's overriding sensibility, opting instead for the notion that anything, in and of itself, is worthy content" (78).

By choosing the Wild West as the motivating context, I do not mean to imply an absence of social or moral structure. Structure is there in the form of rules, regulations, and sheer technical requirements for operation of the technology. Online services and Internet Service Providers (marshalls and sheriffs) such as America Online or Geocities impose laws within their own domains (towns and territories) just as newsgroup moderators regulate which messages may appear and which must get out of town by sunset. The West does not mean isolation: McLuhan and Fiore call attention to the collapse of "older, traditional ideas of private, isolated thoughts and actions" in today's new context of "instantaneous electric information retrieval" (12). That instantaneous retrieval creates a connectedness among "the concerns of all other men," in their view (16). Similarly, scholars such as David Tomas (34), Sven Birkerts (205), and Benjamin Woolley (7) echo McLuhan's own concept of the "global village" or a "net" as a form of connectedness enabled by electrotech.

But this village is a web of connection as open, loose, free, and decentralized as was the West. George Landow describes "hypertext," his way of expressing the idea of cyberspace, in these terms:

> One of the fundamental characteristics of hypertext is that it is composed of bodies of linked texts that have no primary axis of organization. . . . Although this absence of a center can create problems for the reader and the writer, it also means that anyone who uses hypertext makes his or her own interests the de facto organizing principle (or center) for the investigation at the moment. (11–12)

Mark Poster also stresses the decentralized nature of the Internet, in which people generate their own messages and agendas, against which he contrasts the regulatory instincts of both government and big business. Ullman's description of the software engineer's world sounds like Dodge City: "We live in a contest of the fittest, where the most knowledgeable and skillful win and the rest are discarded" (146). It is that vision of cowboy, semi-renegade hackers riding where they will, often on the edges of the law, that captures the sense of electrotech's context as the Wild West.

The ability of electrotech to put power on the human scale into the hands of decentralized individuals is the other side of the global village; it is the cowboy countervalence to frontier justice and the law west of the Pecos. The apostle of the global village, McLuhan, likewise acowledges that even as it connects, "electricity does not centralize but decentralizes" (47). Donna J. Haraway stresses this cowboy side of the global village, in which "the new technologies seem deeply involved in the forms of 'privatization.' . . . Technologies like video games and highly miniaturized televisions seem crucial to production of modern forms of 'private life' " (168).

RHETORICAL POTENTIAL FOR ELECTROTECH

As with mechtech, the potential for rhetorical application immanent within the dimensions of electrotech aesthetics should be clear by now. Electrotech has served many rhetorical purposes in the past. And as with mechtech, the aesthetics of electrotech have facilitated ambivalent reactions: it is the look and feel of a future that is both prosperous and threatening. Electrotech has been both welcomed and feared, rushed toward and run from.

Thomas Hine has studied the ambivalent meanings facilitated by electrotech during the period 1954–1964, in which popular representations "embraced the rocket and the covered wagon with equal fervor" and "the look of motion and efficient technology was balanced and mitigated by the look of togetherness. . . . The kitchen was futuristic, the dining room Early American" (8). More recently, Robins has called for consideration of the social and political effects of cyberspace, treating it as an aesthetic experience with troubling rhetorical potency (152). On the other hand, Paul Crowther argues that " 'Hi-tech' is invested with glamour and high social status for those who are shown to possess it" (12). Yet sometimes the electrotech industrial base seems to possess its users, creating the potential fear of job and identity loss; according to Ewen, "The automated workplace has, for many who must march to the digital beat, led to a failing sense of self" (187). In this section, we will study two rhetorical uses of electrotech that are very much in evidence today: utopian fantasies and the appeal of cybernetics.

Utopian Fantasies

The ability of electrotech to create a cyberspace, an inner depth governed by the imagination and mental capacities of its explorers, makes it especially conducive to the creation of utopian fantasies. Utopias are inherently rhetorical, of course, creating as they do standards against which real societies and actual lives may be compared. A cyberutopia may be *especially* appealing in that its simulations seem real and thus achievable.

Robins is explicit in calling this ability to create virtual worlds utopian, and claims that "all this is driven by a feverish belief in transcendence; a faith that this time around, a new technology will finally and truly deliver us from the limitations and frustrations of this imperfect world" (136). Whereas mechtech fueled utopian visions of a future society founded upon its promises, electrotech utopias are envisioned as achievable now, within cyberspace, at the speed of technological innovation if not of light. Cyberutopia may thus add a rhetorical appeal of urgency or immediacy to the usual social prescriptions implied by utopias.

In other words, electrotech utopia is as near as the closest Game Boy or Netscape Browser. Here we don "spectacular bodies" (Clark, 125) to explore mysterious landscapes or battle thrilling monsters. The world created there is wonderful, as Robins notes: "Existence in cyberspace—a space in which real selves and situations are in suspension—encourages the sense of identification and symmetry among individuals" (150). Note the rhetorical potency in the "real" world of that vision of "identification and symmetry," speaking as it does for equality and community.

Cyberspace may speak to one's real life problems rhetorically as well. Turkle argues that "for many people the computer at home becomes a tool that compensates for the ravages of the machine at work" (*Second*, 170), a truth evident to anyone who has relaxed in the evening by cruising the net or exchanging e-mail. The potential of cyberspace for creating powerful simulations for users is a rhetorical resource; as Poster argues, "Virtual reality machines should be able to allow the participant to enter imagined worlds with convincing verisimilitude, releasing immense potentials for fantasy, self-discovery and self-construction" (93). The lessons learned in a cyberutopia may be applied to rhetorical exigencies in real life—for better or worse, realistically or unrealistically.

Electrotech has had rhetorical implications for utopia beyond cyberspace, however. The electrotech aesthetic means newness and innovation. As the look of the future, it may urge what the present should be so that we may achieve that future. Nye notes the way in which increasing electrification during the Great Depression oriented "public discourse toward an ideal, technological future" (351). Such a reorientation would be of great rhetorical value in gaining adherence to social policies. This futuristic utopian application also was true in the early days of electrotech, with its transitional aesthetic expression in streamlining. Jeffrey L. Meikle calls attention to the "clean-lined" aesthetic of streamlining, and that incorporation of that design feature into appliances and household goods "prepared the way for public acceptance of streamlining as a style that promised to eliminate complexity and friction from society in general" (186), beyond mere design. This strategy, Meikle notes, was a rhetorical response to "a common assumption that society's larger processes had to be rendered smoother, less complex, more frictionless in operation. Streamlining . . . visually expressed this utopian desire" (189). When we consider the labor unrest and class divisions that marked American society as streamlining was being introduced, we can understand how that aesthetic served hegemonic goals of creating order and stability.

We noted in chapter two how mechtech aesthetics came to be loaded with rhetorically powerful morality. Writing in the 1930s, Cheney and Cheney also note the moral weighting attached to streamlining in their remarkable equation of the airplane (the inspiration for streamlining) and the cross: "The machine-conscious mind begins to relate all such products of scientist-artist design back to the most conspicuous symbol and inspiration of the age [the airplane], as the reverent medieval mind related everything to the symbol of the cross" (102). They equate a streamlining aesthetic with life itself in claiming that "we subjectively accept the streamline as valid symbol for the contemporary life flow" (98). These equations rhetorically position electrotech aesthetics as signs of a utopian future.

The Appeal of Cybernetics

Cybernetics refers to a wide-ranging problematic, founded by the work of Norbert Wiener, of the intersection between people and machines. As Featherstone and Burrows explain, "Cybernetics encompassed the human mind, the human body, and the world of automatic machines and attempted to reduce all three to the common denominator of control and communication" (2). It includes pacemakers and hearing aids as well as Robocop and the Bionic Woman. Cybernetics incorporates media images and fictional accounts of robots, news stories about automation and artificial intelligence, and social commentary and philosophy about the merger of humanity and technology. In the next chapter we will see how fears about technology are often expressed through the aesthetics of the decayed machine, including images of the evil and uncontrollable cyborg. In this chapter, we consider the positive side: rhetorical uses that have been made of cybernetics as a positive image, employing electrotech aesthetics as signs of hope, progress, and perfection.

The image of the cyborg, "a self-regulating human-machine system" (Featherstone and Burrows, 2), a merging of flesh and machine, is central to the rhetoric of electrotech aesthetics. The mechanical person did not first appear with electrotech; James J. Sheehan and Morton Sosna explore historical equations and overlaps between people and machines. Allen Newell argues that although "the computer is a metaphor for the mind . . . Western society has always used the technologies of the time as metaphors for the human mind" (159). Mumford observed that "more and more, from the sixteenth century on, modern man patterned himself upon the machine" (*Art*, 12). But the escalation implied in Mumford's statement indicates that with electrotech that conjunction has reached a new peak, with electrotechnology perfecting the ways in which humans and machines may merge. Nye makes this explicit: "People do not merely use electricity. Rather, the self and the electrified world have intertwined" (390).

Electrotech aesthetics make machines seem more human, an equation which can rhetorically support the argument that humans are machines. Samuel Florman identifies this important shift in the human-machine equation as being brought about by electrotechnology: "And as the 'mechanical' machines, which remind us of muscles, bones, and circulatory systems, are supplanted by an increasing number of electronic and chemical devices, which imitate the processes of the brain and nervous system, the sense of vitality in the machine increases" (139). Lupton agrees: "The cultural meanings around PCs, including common marketing strategies to sell them and the ways people tend to think about their own PC, relies on a degree of anthropomorphism that is found with few other technological artifacts" (98). Mumford foresaw this merger in noting the ways in which "humanizing the machine" was leading to "mechanizing humanity" (*Art*, 5).

Today, the image of the cyborg is widespread. As Haraway put it, "By the late twentieth century, our time, a mythic time, we are all chimeras, theorized and fabricated hybrids of machine and organism; in short, we are cyborgs" (150). One might just as well say that electric machines are people, and thus cyborgs; Jean Baudrillard does so: "Because the automated object 'works by itself,' its resemblance to the autonomous human being is unmistakable" (*System*, 111).

Let us consider what facilitates this merger of the machine and human. At a very basic level, the fact that machines are human creations invests some of the human in machines, and vice versa. As W. H. Mayall put it, "When we look at any man made artifact, we also look through it to its own creator" (89, 93).

The dimensionality of electrotech also facilitates cybernetic union, in that one enters the depths of cyberspace by cognitively changing and joining the machine. Virtual reality, Larijani claims, "aims to dim the line between the machine and the thinker outside it" (9). When the electrotech user enters the depths of cyberspace, in at least an aesthetic sense she has merged her humanity with the machine and become a cyborg; Turkle describes the process: "When hackers are inspired and in creative telepathy with the machine, it is as if they are inhabited by the medium, inspired by the muse" (*Second*, 228–29). That result, too, was foreseen by Mumford: "To succeed in operating machines he must himself become a subsidiary machine" (*Art*, 46).

Another way in which electrotech dimensionality facilitates cybernetics lies in the way electrotech confounds traditional borders and boundaries. Whether one is actively engaged with cyberspace or not, the distinctions between electrotech objects on the one hand and humans and our world on the other are blurred. Haraway says so explicitly: "In the traditions of 'Western' science and politics . . . the relation between organism and machine has been a border war" (150). This border war, she maintains, has "made thoroughly ambiguous the difference between natural and artificial, mind and body, self-developing and externally designed, and many other distinctions that used to apply to organisms and machines" (152). Featherstone and Burrows note the border-dissolving effect of cybernetics:

> The key analytical categories we have long used to structure our world, which derive from the fundamental division between technology and nature, are in danger of dissolving; the categories of the biological, the technological, the natural, the artificial *and* the human—are now beginning to blur." (3)

Joseph Tabbi similarly notes our increasing merger of human and machine, and argues that this blurring of boundaries is achieved aesthetically:

> One of the more radical and surprising responses to this disturbing situation has been not to separate ourselves further from the machine, but to make technological mechanism the site of an aes-

thetic embodiment where mind and world are neither opposed nor merged. (21)

Electrotech's erotic of the mind is another facilitator of the cyborg aesthetic, for it is on the shared basis of thought or thought-like processes that the cyborg merger is often founded. Tomas argues that machines and humans are merged, not on the basis of any external resemblances, but on the basis of a shared dependence on and use of *control*: human communication processes and feedback mechanisms in machines both allow for control over and adjustment of one's actions and purposes (27). This means that people have an affinity for the cyborg that creates a ready resource for rhetorical exploitation: its processes are our processes. McLuhan makes the same argument from his perspective of communication, claiming that electrotechnology is making humans into cyborgs: "In this electric age we see ourselves being translated more and more into the form of information, moving toward the technological extension of consciousness" (64).

Those several bases for the aesthetic union between people and electric machines, between our world and cyberspace, create tremendous rhetorical potential. If the person = the cyborg or our world = cyberspace, then assertions may be made for either side of the equation about how that side should be defined and treated. As Kenneth Burke would say, this assertion of a factual "is" implies a rhetorical "ought."

Much of the rhetorical use of the images of the cyborg exploit this human/machine link. Although the equation might be stressed in both directions, Tomas argues that "cybernetics operationalized the question of 'life' by displacing the concept of organism from biology to engineering, thus effectively transforming it into a hardware problem" (26). To see people as part machine has rhetorical impact; at the very least, it implies how people ought to be treated. Furthermore, if one appreciates machines, one may see the human as a beautiful machine. As Baudrillard puts it, "Man . . . by automating his objects . . . reveals in a way what part he himself plays in a technical society: that of the most beautiful all-purpose object, that of an instrumental model" (*System*, 112). Felicia Miller Frank suggests another rhetorical use of exploiting the cyborg: images of "the artificial woman," she claims, have long been a site of struggle in gender politics. That image in literature has been used to define women, as she shows in her study.

The boundary problems posed by the cyborg, as noted above, are exploitable for rhetorical purposes; machines may be discussed in human terms, humans in machine terms, because of the eminence of the cyborg in popular consciousness. Gregory Benford notes that very thing in claiming that "the increasing connection of ourselves to our machines is unsettling and ripe for mingled scientific imagery" with rhetorical potency (227). As Haraway argues, the "cyborg myth is about transgressed boundaries, potent fusions, and dangerous possibilities which progressive people might explore as one part of needed political work" (154). Machines may be thought of as people, as in Lanham's claim that the computer "is now a personal companion rather than an impersonal giant" (ix). Discourse about the machine may thus be a metaphor for discourse about human problems, and vice versa. Tabbi argues that the recent works of Pynchon, DeLillo, and others employ this rhetorical strategy: "When we contemplate technology in these works, we are ultimately contemplating ourselves, for the expressive potential within the machine returns us to the source of all thought, the human mind in relation to other minds" (20).

Specific aesthetic dimensions of the cyborg merger have rhetorical potential. Featherstone and Burrows' lyrical description of entering cyberspace is cast in aesthetic terms: "Once in the matrix, operators can 'fly' to any part of the vast, three-dimensional system of data coded into various colourful iconic architectural forms laid out beneath them like a vast metropolis: a city of data" (6). Note the rhetorical uses of situating the user as a cyborg, an electronic creature of the matrix, in a *city*, with possible connotations of civic responsibility, belonging, and community. Tomas argues that early depictions of cyborgs in popular culture represented them as free from normal physical constraints, thus linking electrotech to a discourse of freedom and possibility with clear rhetorical implications (35–36).

CONCLUSION

This brief review of the rhetorical uses of electrotech in utopia and in cyberspace is far from exhaustive. Other rhetorical applications are supported by the potential inherent within electrotech aesthetics. We must also remember that electrotech aesthetics may be facilitated by texts that also support mechtech readings. In other words, mixed modes of aesthetic experience are not unusual, and thus provide the resources for rich rhetorical applications. Mixed aesthetic modes are not surprising given the historical relationship between mechtech and electrotech, in which the latter in many ways grew out of the former.

Mechtech and electrotech are also linked in that both tend to treat the machine positively. Even when the aesthetic experience of either casts the machine in a sinister light, the texts tend to treat the machines in straightforward, literal ways. In the next chapter, we turn to our third and final type of machine aesthetic, *chaotech*, or the appeal of the decayed machine. We will see that chaotech in many ways is an ironic reaction to or turning of the aesthetics of mechtech and electrotech. We have seen the literal ground for chaotech in chapters two and three; now we turn to the ways in which it works off that ground.

CHAPTER FOUR

Chaotech: Aesthetics of the Decayed Machine

There is a category of aesthetic experience of machines that is qualitatively different from the others. I refer to the aesthetic appeal of decayed machines. This is the sensibility that finds artistic joy in junk yards, disemboweled televisions, and automobile graveyards. It is the sensibility of the young person enjoying the "Industrial" scene in dark clubs, listening to music inspired by the sounds of factories. This is the aesthetic of dark visions of science fiction, filled with dysfunctional machines cobbled together from odd parts of dying computers and engines.

Chaotech, or the aesthetic of the decayed machine, is distinctive in not taking machines on their own original terms. Instead, it experiences machines as changed, as something other than what they originally were. In this sense chaotech is *ironic*, keeping a knowing distance from its object, deriving pleasure from understanding what a machine once did and once was but is and does no longer. In this chapter we will first consider evidence indicating that there really is an aesthetic of the decayed machine and we will look for some root causes and contexts for that aesthetic. Then we will use our nine dimensions of machine aesthetics to understand what constitutes chaotech.

THE APPEAL OF RUINS, DESTRUCTION, AND CHAOS

> The current fashion for "happenings" has brought the great science-fiction event of the "suicide" or murder of the object a little closer to home. The happening involves an orgiastic destruction and debasement of objects, a veritable hecatomb whereby our whole satiated culture revels in its own degradation and death. A recent fad in the United States amounts to a mass-marketing of the happening in the shape of novel contraptions, composed of gears, rods, shafts and what-have-you—true jewels of useless functionality whose merit lies in the fact that they fall apart of their own accord, suddenly and irreparably, after a few hours of operation. (Baudrillard, *System*, 122–23)

Jean Baudrillard's specific example must call to mind certain advertisements on late night minor television stations, and the gadgets for cook and handyperson that are sold there. But his general point refers to a cultural fascination with the machine gone wrong. As he also notes, "The theme of the robot that goes off the rails and destroys itself is a common one. . . . There is a secret apocalypse of objects—or of the Object—which fuels the passionate interest of the reader" (*System*, 122). Baudrillard envisions an aesthetic love in postmodern culture not for the occasional malfunction but for a landslide of mechanical disaster, a whole industrial complex crashing down at once: "A technical hitch infuriates us, but an avalanche of technical hitches can fill us with glee" (*System*, 131).

Baudrillard is not the only cultural observer to note a fascination with the decayed machine. Evidence of that aesthetic is all around us. Ellen Ullman, a software engineer, claims that "there's a perverse comfort in broken machinery" (65). Manfred Hamm's book, *Dead Tech: A Guide to the Archaeology of Tomorrow* is full of photographs of ruined military machines, auto graveyards, abandoned factories, and other images of chaotech. They are presented, if not as art, then as objects for artistic contemplation. Michael Selzer identifies a similar fascination with ruin in Nazi aesthetics: "The Nazis . . . not only built neoclassical structures intended to last for a thousand years, but designed them, in accordance with Albert Speer's 'theory of ruin value,' so that they would look imposing even when they eventually collapsed into ruins" (5). Selzer goes on to locate that love of ruin and waste in our culture not in grand buildings but in machines: "No Ozymandiases we, our model is the plutocrat who scraps his Rolls-Royce when it has run out of gas" (6). Paul Zucker notes that in all fascination with decay and ruin there remains an "aesthetic unity" of the experience that moves it beyond a mere observation of simple collapse and decay and into the realm of art (3).

A fascination with decay has cultural antecedents going back at least to the Gothic Romantic movement of nineteenth century Europe. Adrian Del Caro notes the centrality of a Dionysian aesthetic of destruction in the work of Friedrich Nietzsche, featuring "annihilation and the pleasure derived from ruination" (23). That sensibility is found beyond Nietzsche, of course; it is the feeling that "by engaging in the process of annihilation, i.e., by participating in becoming Dionysian, happiness is experienced. The old is not mourned, because out of necessity it must yield in the flux of chaotic forces" (Del Caro, 18). Images of decay embodied in the ruined abbey or tumbled down castle are found in the works of pre-Raphaelite and other Romantic artists; that same sensibility was, then, culturally available for application to machines as well as to monasteries. And Walter Fogg notes the late nineteenth century fascination with "disasters connected with man-machine relationships," as in the novel *Frankenstein*, as a precursor of our current distrust of technology and machines. In sum, the idea that a ruin or specifically a ruined machine can be an aesthetic experience is solidly established in our cultural history. But why, and what social, psychological, economic, historical, or other causes underlie today's aesthetic of decayed machines?

SOME SOURCES OF CHAOTECH

What gives rise to a sensibility for the decayed machine? A number of answers might be given. Several scholars have given *psychological* answers to the question of how a fascination for chaos and collapse arises. Morse Peckham observes that "man is inflamed by a rage for order, and the more he fails in controlling his transactions with his environment, the more the flames of that rage consume him" (*Man's*, 34). Yet such failure of order and control is inevitable. The essence of art is not that it imposes order on subject matter, Peckham argues, since order is a component of the act of experiencing and perceiving rather than of the object of art itself (33). Instead, art is a "rehearsal" for the chaos that must inevitably come to all of us, even as we rage for order (313–14). Elsewhere, Peckham argues in a related way that art produces disorder in the perceiver so as to also restore a wider sense of order, producing a satisfying flirtation with chaos (*Art*, 100). This theory suggests that in appreciating the decayed machine we are practicing for acceptance of the unavoidable decay that will come to our possessions and our bodies.

Other scholars have pointed to order and disorder as opposite sides of the same coin, suggesting that a psychological explanation for chaotech lies in its implication that if there is disorder and decay there must also be order and life; the one is an index of the other. Rudolf Arnheim, for instance, notes that "order is a necessary condition for anything the human mind is to understand" (1) and that "man imposes orderliness on his activities because it is so useful, cognitively and technologically, in a society, a household, a discourse, or [NB] a machine" (34). But Arnheim also turns to Freudian theory to argue that tension reduction is a human imperative as well, a motive served by art (44–45). Aesthetic experience is not so much the experience of order as it is a reduction of tension, some of which may actually come from the constraints of creating and maintaining order. Because any "organized structure will simply succumb to disintegration, either by corrosion and friction or by the mere incapacity to hold together," one form of aesthetic experience is the reduction of tension by embracing images of that disintegration. Arnheim points to examples that come close to chaotech, an exhibit of "deflated giant models of orderly functional instruments" (52–53). Chaotech may appeal psychologically, then, as an experience of tension reduction, a momentary respite from the rigors of maintaining an orderly world.

Similarly, John Kouwenhoven notes that although "there is, or seems to be, something about the human mind that is affronted by chaos" (224), on the other hand "we must . . . conceive of existence as some sort of process maintaining an unstable equilibrium between contrary tendencies" of growth and decay, life and death (225). Applying this psychology to art, Kouwenhoven argues that "all the perfectly patterned relationships the artist seeks to symbolize in design are in an ultimate sense untrue and, in a profound sense, antilife" (223). Life includes chaos and death as well, and art can sometimes celebrate that reality as well. Extending this idea to our concerns, we might understand chaotech as an aesthetic experience of the death and decay side of that "unstable equilibrium," in which the decayed machine appeals as the other, inevitable side of the new, gleaming, functional machine. Chaotech is a balancing aesthetic, in this view, between what Selzer calls "the simultaneous passions—each requiring the other—for destruction and renewal which infuse so much of the modern world" (6).

In contrast to psychological explanations of the sources of chaotech, *sociological* explanations have been advanced as well. It has become commonplace to observe that many people live in a "postmodern" world of fragmentation and disconnection, brought about by social, economic, and technological developments. Patrick Novotny argues that postmodern thinking is essentially parodic, and that "this parody transgresses aesthetic style and representational norms, and celebrates the fragmented, indeterminate, and unpredictable subject" (99). We might extend that observation to claim that chaotech is a parodic celebration of the fragmented machine, even of the broken machine as an image of the broken society, the broken person. Chaotech is an aesthetic of our time in its celebration of disconnection and dissolution.

Several observers suggest that a loss of faith in technology underlies the aesthetic experience of chaotech. Stuart and Elizabeth Ewen identify serious doubts concerning technology from the beginning of the twentieth century or before: "The fantasy of a smoothly running mechanical world was eroded by the dislocation and unrest that it provoked" (13). Lewis Mumford would observe in the first third of the twentieth century that "the machine is ambivalent. It is both an instrument of liberation and one of repression" (*Technics*, 283), such that "unquestioned faith in the machine has been severely shaken" (365). Today, Ignasi de Solà-Morales Rubió and Josep Ramoneda observe a failure of confidence in many practices and institutions, including technology:

> After years of cold war the world is now undergoing a difficult and risky reorganization. Astrophysics, genetics, and psychology provide no guarantees. Technology does not appear to be a sure answer, and politics seems unable to avoid authoritarianism, corruption, and constant conflict. (20)

Walter Fogg agrees:

> Despite the early optimism about technology and social progress, there is strong opinion in our century, expressed in contemporary dystopias and science fiction, that technology is not only beyond the control of society but that man is still largely unconscious of the profound revolution he is bringing about. (60)

Even an engineer such as Samuel Florman notes gloomily that "engineering's Golden Age ended abruptly about 1950, and . . . the profession, for all its continuing technical achievements, finds itself at the present time in a Dark Age of the spirit" (11). Chaotech may well be understood as an expression of that dark spirit not only among engineers but among the general population.

Another sociological explanation for the roots of chaotech sees its aesthetic as an expression of urban problems. Cities and machines have been long connected, of course, as urban areas have been sites of industry, invention, science, and technological innovation. Urban decay is often manifested in the decay of machines, as factories sit idle and rusting, outdated machinery and landfills collect the detritus of dead automobiles, washing machines, and toasters. Preoccupations with the dead machine reflect an engrossment with the dead city.

Frances Bonner vigorously sums up the grim cityscape for which chaotech is a mirror:

> Civilization has peaked, and while there still may be new scientific and technological developments, these do not equate with progress. There are no pretensions to egalitarianism, and for most people life is already hell. The convention used to convey this is remarkably similar in literature, film, and TV programming—the run-down inner-city slum-cum-tent settlement, overcrowded, trashed and graffiti-ridden. (193–94)

In Bonner's view, the decayed city is a representation of a decayed industrial culture generally. Paul Crowther argues that cultural experience is coming increasingly to address problems arising from urban and industrial collapse: "Experience in a modern society is oriented fundamentally towards the containment of shocks arising from an abundance of adverse stimuli in the industrial work process and big-city environment" (11). An aesthetic of decayed industrial machines will then be grounded in and make much use of images of the blighted city, all of it representing the perception of a collapsed worldwide industrial culture.

The collapse of the city is not only physical, but social as well. Solà-Morales Rubió and Ramoneda write of social, communicative as well as physical dysfunction:

> The cities, the site of public and private activities, are subject to constant action, the consequences of which are not positive. The physical construction of environments and the symbolic structure through which collective communication is produced are today

developed in erratic ways, with deceptive ideological contents and responses that experience has already proved to be invalid. (20)

One can imagine how poorly constructed, failing machines might come to represent the clattering, blocked malfunction of the city. Craig J. Saper suggests that urban decay has to do with attitudes as well as actual physical decay; it "depends on more than loose bricks or garbage on the sidewalk. It depends on a more general 'sense of a situation' " (105). These authors, then, are in agreement that urban collapse is a matter of social perceptions as well as actual decay. That means that it is also an aesthetic experience, and that chaotech may be one of the most important and widespread expressions of social reactions to changing urban environments.

Many sources of chaotech aesthetics have been suggested. In the following review of the dimensions of chaotech, we will keep in mind its ironic nature. The chaotech experience always takes into consideration what the decayed machine was or might have been. Chaotech is an aesthetic of possibilities lost or abandoned; it appreciates its objects both for what they are now and for what they were. We will also bear in mind the potential of chaotech to mirror a postmodern culture in fragmentation and disarray.

DIMENSIONALITY: THE BREACHED HULL

Chaotech's experience of the surface and interior of the machine confounds the distinctions and tensions of mechtech and electrotech. The decayed machine has lost its *integrity*; its guts have come spilling out. One is aware of the surface of the machine, yet it has been ruptured, corroded, rusted, torn open. One is aware of the interior parts of the machine, yet they have fallen out, shorted, seized up, melted solid, collapsed upon themselves. The dimensionality of chaotech is an experience of shocking openness. Yet it can also be pleasing; Ullmann celebrates the breached hull in declaring, "I like the look of delicate circuit boards open to the naked air" (65). An internal process, whether mechanical or electrical, that once ran smoothly within its shiny shell has been disrupted by an invasion of air, water, earth, and fire. The paradigm example of chaotech dimensionality is the breached hull of a drifting space ship, the electrotech and mechtech interior dark and still in the vacuum of space.

A dimensionality of the breached hull can be experienced as a disjunction of modes or dimensions. Surface and depth, nature and technology, human and machine—these and other tidy distinctions are confounded in chaotech. Sherry Turkle points out that "the computer, standing between the physical and the psychological, creates a new disorder" (*Second*, 32). The disorder of which she speaks is one of perceived incommensurability between the body and the mind, a stock distinction that computers seem to muddle. One does not have to experience electrotech machines in that negative way, of course, but when one does the sensibility of parts that do not belong together could be expressed in a chaotech aesthetic of parts that have decayed so much that their integral, functional connection is broken. In other words, if chaotech is experienced in part as a disjunction of modes or dimensions, then some revulsion toward workable mechtech and electrotech machines may be expressed as a chaotech aesthetic.

Another disjunction is suggested by Paul Zucker: "Devastated by time or by willful destruction, incomplete as they are, ruins represent a combination of created, man-made forms and organic nature" (3). Although he is speaking of architectural decay, Zucker's point is applicable to the decayed machine. Although any made object may be said to confound a distinction between human and organic forces, functioning mechtech and electrotech machines may be said to represent the (temporary) triumph of human design and intention. A destroyed machine represents the assault of time and the elements, the organic. It is neither clearly a human creation nor a product of nature alone. In that sense, chaotech dimensionality confounds many other dimensions as it confounds the dimensions of surface and depth.

SUBJECT: NOSTALGIA

The subjective "use" of a decayed machine is different from the mechtech or chaotech experiences. The original functionality is no longer possible to achieve, so one cannot use the machine in the sense of making it "work." The subjective experience of chaotech must therefore apprehend the machine as something linked to the self; we identify with the ruin. In that sense, the subjective experience of chaotech is one of *nostalgia*. We experience the decayed machine for what it *was* and used to be able to do. Much of that experience may involve memories of what subjects themselves once were and once used to do with the machine: the sight of a rusted hulk of a '57 Chevy may remind them of their younger, hot-rodding days—discovery of a burned out Tandy computer in the basement may take them back to their early days of computing. Chaotech subjectivity thus links nostalgia for the machine's lost past to nostalgia for one's *own* lost past. Florman's poignant observation that "the unpleasant truth is that today's engineers appear to be a drab lot" seems to reflect such a nostalgia (93)—perhaps because today's engineering is often perceived through a chaotech aesthetic.

Zucker has noted this nostalgic dimension in the memories evoked by ruins: "Although the ruin still continues to exist in the sphere of life, life has departed from it, and we are aware only of a more or less well-preserved fragment of an earlier age" (2). It may be more accurate to say that the old life has departed from the decayed machine; the new relationship, especially the new aesthetic relationship, into which it enters with the perceiver, is a new sort of life. Thus, Zucker also emphasizes the relationship between the ruin and the person as a prime determinant of how it is perceived (2). With a ruined machine as well, we rarely experience it only for what it is: a collation of nonfunctional metal and plastic. It is also experienced for what it was in the experience of the perceiver, and is thus aesthetically nostalgic. In this sense, chaotech makes use of *detournement*, as explained by Novotny: "The works of the postmodern aesthetic have often relied on the technique of detournement. . . . Detournement is the appropriation of existing cultural fragments in such a way as to alter and invert their meaning" (100). Subjective chaotech aesthetics is this sort of appropriation of ruined machines for nostalgic purposes, taking nonfunctioning pieces of what used to be a machine and constructing for it a past linked with the perceiver.

OBJECT

The chaotech object is an intriguing, macabre spectacle spanning a range from contemplation of quiet decay to glorification of apocalyptic vistas. The chaotech object offers an aesthetic of rust, corruption, dirt, waste, chaos, and entropy. It is the detritus of a closed, obsolete factory, the shorted circuitry of an ancient motherboard, the smashed watch. A paradigm of chaotech objects may be found in any of the *Mad Max* films' postapocalyptic visions, which create scenery, costumes, and landscape from broken bits and pieces of mechanical and electronic debris.

Perhaps the overriding aesthetic component of the chaotech object is, of course, the fact of decay itself. As Zucker observes, "A ruin exists in a state of continual transition caused by natural deterioration, specific catastrophes, or other circumstances" (2). That "transition" is the process of decay, or becoming ruined, whether for an abbey or a calculator. In this sense the chaotech object is an exemplar of the postmodern condition of the society from which it springs as well. Novotny reminds us that fragmentation and the collapse of received structures is of the postmodern essence (100–1). Fears and anxieties about that cultural process might then be embodied in the chaotech aesthetic object and in the deterioration that it experiences.

Cityscapes are often perceived as chaotech objects, and artistic representations of cityscapes sometimes emphasize their desolation so as to potentiate that effect. Fredric Jameson, discussing apocalyptic visions, points to the example of "the high-rises of Beijing itself, which equally seem to fall out of the sky and randomly to people the hitherto empty terrain around the city loop in numbers of housing units so great as to seem meaningless" (41). High rise apartments are neither failed machines nor broken electronic devices, yet I think we can see the chaotech aesthetic of collapsed order, process yielding to randomness, and efficiency giving way to clutter in Jameson's not-unusual view of a large, corrupt city. That chaotech aesthetic also underlies Jameson's example of "the dystopian imagination: the image of the mortal totality of a city in flames given over to destruction" (41). That aesthetic may be found in several films such as *Blade Runner*, described by Jenny Wolmark in this way: "Dekker's world-weary voice-over evokes a . . . moral alienation from futuristic city streets that are a chaotic jumble of conflicting messages and signs" (113).

Wolmark's reference to disorder in both city and signs recalls our earlier consideration of chaotech as a postmodern aesthetic expressing the fragmentation and disunity of life today. Saper observes that "urban decay is the *jouissance* [destabilizing] of architecture and modernity" (107)—in other words, the broken city mirrors and represents not merely buildings but the cultural condition. Saper argues that because so many of the problems of our culture are linked to the city, fears about industrial, urban culture are played out in images of the city's decay: "The birth of the modern city created not only peculiarly urban problems but also the motivating force linking a noir sensibility with urban desolation, malaise, and destabilization" (108). Saper usefully outlines the dimensions of that urban "noir sensibility":

> Those elements include the seduction of the seamy, the pleasures of watching decay as a changing, evolving force rather than as an end in itself, and the hysteria of living in a contemporary city as almost pleasurable in its unpredictability. (109)

We may take Saper's list as a handy summation of the aesthetics of the city when perceived as a chaotech object.

We have observed before how chaotech offers an ironic commentary on mechtech and electrotech. In the second chapter, we noted how Bauhaus design and architecture came to reflect a mechtech sensibility of orderly duplication, uniformity, and efficiency. A chaotech read of Bauhaus interprets the orderly, machine-like rhythm of housing units in the urban context as horrible and corrupted. Josep Ramoneda perceives Bauhaus housing developments in Dessau, Germany in this way: "The infinite array of matchboxlike buildings packed together like honeycomb, housing thousands of people in a climate that totally precluded uniqueness and the pleasures of living"(54).

PRODUCTION

The chaotech aesthetic of production is in reality an aesthetic of no production at all, a joy of gridlock and dysfunction. In keeping with chaotech's ironic stance, such an aesthetic must always find its place in relationship to what a machine once did or is supposed to do. The aesthetic also demands functional failure on the part of the machine in its current state. This is not an aesthetic of recycling: the resurrected machine or mechanical part is reborn and has no place in this aesthetic. Zucker puts it succinctly in discussing ruins when he says that "functional values which the ruin might have possessed originally are of even less value in its aesthetic interpretation. If structurally adapted to the needs of later centuries, ruins lose their character as ruins" (2). The paradigm case of chaotech production aesthetic might thus be the automobile graveyard with its rows upon rows of nonfunctioning machines, their power to run locked up within their seized and rusted gears.

Novotny finds this aesthetic of failed production in his description of factories within the decayed cybertech cityscape: "The smokestacks with the fumes from an earlier era of industrial production yield to a world in environmental breakdown, a perpetual haze of smoke and filth with constant rain streaking the concrete on a landscape of abandoned industrial factories" (104–5). In that vision, the earth itself becomes a kind of failed machine, in "breakdown," that parallels the collapse of the factories that have spoiled it.

We might think of the chaotech aesthetic of production as an expression of postmodern fragmentation, as a cultural symptom. The decayed machine is thus a machine that refuses for one reason or another to do what the modern industrial economy says it must do. The center has failed to hold for the broken machine, as it has failed for the economy and the society as a whole. In this sense I note an interesting if odd parallel discussion of ironing by Elizabeth Diller. She argues that in the act of ironing "the shirt is disciplined, at every stage, to conform to an unspoken social contract" (156)—and we might recall the close connection between mechtech machines, at least, and a vision of a rigidly ordered society. Diller then reviews an art museum display of aesthetically crumpled shirts, speculating, "What if the task of ironing were to free itself from the aesthetics of efficiency altogether? Perhaps housework could more aptly represent the postindustrial body by trading the image of the *functional* for new imperatives of the *dysfunctional*" (156–57). If we but substitute "machine" for ironing or housework, we have a vision of the decayed machine as a rejection of industrial discipline, and thus as a celebration of postmodern entropy.

George H. Marcus calls our attention to the oft noted replacement of functionalism as a foundation of design with a jumble of styles in postmodernism (*Functionalist*, 153). He claims, "The core works of Breuer, Mies, and Le Corbusier now fill a niche as one of the many historical styles that may be selected for interior decoration, totally removed from their independent aesthetic associations" (160), a decenteredness and freedom he describes as "a melange that would have been unthinkable two decades ago" (162). Might we fruitfully think of a confused, centrifugal melange, a lack of direction or anchor, as not only the postmodern condition but as the central characteristic, the "problem" if you will, of the decayed machine? Its parts no longer work together. If postmodernism "celebrates the fragmented, indeterminate, and unpredictable subject," as Novotny puts it (99), might that celebration be expressed in an aesthetic of the fragmented, indeterminate, and unpredictable machine—the machine that can no longer produce? The links are snapped, the tie rods broken. Were the engine to start its parts would flap around uselessly rather than in sync. Were the switch to be turned on, useless circuits would channel electrons to no apparent purpose.

GENDER: THE TOON

A toon, of course, is colloquial for *cartoon*, or more particularly for a character in one. The term first came into wide popular parlance with the film *Who Framed Roger Rabbit?* Toons, in that film, were the animated characters who were able, as were the live actors, to cross over from an animated scene to real life. Cast as foils to the actors, they were simulations. The actors represented certain stock character types, and were of course not "playing themselves." In contrast were the toons, who had no life beyond the screen. The toons were usually represented as male or female, and exhibited conventional signs of sexual attraction and behavior; but as toons they were not in fact engendered and could perform none of the usual sexual or reproductive functions typical of "real" males and females. Although in caricaturized ways they resembled real men and women, they did not represent them; their existence was at the mercy of the cartoonist, bounded by the frame of the screen.

In this sense, the aesthetic gendering of the decayed machine is as a toon. A decayed machine carries echoes of the masculine mechtech or androgynous electrotech original, but it is no longer conventionally associated with male or female social identity. The dead hull of a computer may resemble the feminized object it once was, but it is no longer an object of desire, scrutiny, and gaze in the same way. The rusting engine block may echo its masculine past but no longer represents it. Instead, each decayed machine (when perceived as chaotech) is an experience in and of itself: a simulation. To make these wrecks into objects of aesthetic contemplation, which they *need* not be, requires an imaginary act of construction. To see them as aesthetic objects in their own right is to see them as simulations, having meaning and aesthetic value in the here and now and not insofar as they resemble some "real" working machine.

Another way to say this is that the gender identities of toons is rather slim and created out of whole cloth, when a toon has a gender identity at all. The aesthetic existence of the chaotech object is similarly something supported by the imagination of the perceiver. Less able to be used, unable to produce, the chaotech machine has a similar slim existence as a simulation, referencing what it used to be but representing nothing beyond its own facticity.

PERSONA: THE THIEF

We have noted the importance of what a machine used to be and used to do in understanding its appeal as a decayed ruin. The chaotech object is defined in relationship to something else, something whole and productive, from which it steals the premises of its own identity now being constructed by the perceiver. In that way, the chaotech object suggests the persona of the thief. The thief is defined by her relation to others and to objects; she has no other "job" than to appropriate value from those others. As the chaotech object is aesthetically parasitic, so is the thief economically parasitic upon an economic system.

The chaotech machine, like the thief, is also an outlaw object. Machines are not "supposed" to break and collapse, and when they do, conventional aesthetics asks that they be discreetly tossed out. Chaotech operates on the borders of aesthetics, as the thief works on the margins of an economy.

It is worth noting that outlawry is often defined in part through misappropriation of machinery. That misappropriation is not quite the same thing as the destruction or decay of machinery, but it is aesthetically close to it. The mechtech outlaw misuses the gun or the automobile. In *Smokey and the Bandit* we know who the bandit is by his illegal use of a powerful car. We know that Sylvester Stallone's character in the *Rambo* series is at the edge of the law in part through his appropriation of grotesquely large automatic weapons.

Similarly, there is an outlaw persona developed by some computer hackers who choose to work on the boundaries of the law. John Perry Barlow says that computer hackers in the flesh are staid and harmless, "But . . . the boys had developed distinctly showier personae for their rambles through the howling wilderness of cyberspace. Glittering with spikes of binary chrome, they strode past the klieg lights and into the digital distance. There they would be outlaws" (97). Likewise, Katie Hafner and John Markoff refer to the common persona of "the network cowboy, out on the nets having a good time roping computers like so many steers" (123), and we know how close the cowboy is to being an outlaw in popular mythology. The point in sum is that use of machines in outlawed or questionable ways is a kind of thievery, a way to shift both user and machine in the direction of the outlawed and marginal status of chaotech.

DOMINANT RELATIONSHIP: RAPE OR IMPOTENCE

In a sense, the chaotech aesthetic experience is the most purely aesthetic experience of the three sorts studied in this book. With electrotech and mechtech, someone *does* or *can* do something with the machine in question, even when it is regarded purely as an object. But the chaotech experience is purely aesthetic: one regards, perceives, perhaps fantasizes, but one cannot pick up the machine and use it—if one could, it would no longer be decayed. Certainly one cannot use it as "intended," for its original function. In a sense, then, the relationship with the machine is going nowhere, it is at a dead end, it exists only for the moment.

The dominant relationship implied by chaotech aesthetic is therefore either *impotence* or *rape*. The chaotech relationship is impotent precisely because it is nonfunctional. The machinery, in reality or allegorically, does not work. Other forms of contact, whether visual, tactical, or some other form, must replace active engagement. The gaze takes the place of action. Touch stands in for fruitful manipulation. Pieces of circuitry soldered into stained glass compositions, which one may see at summer art fairs, will never again carry electricity: they are simply to be looked at now.

Chaotech may suggest a relationship of rape in recalling the force of destruction and dissolution that led to the current collapsed state of the machine. The machine has been violated, by time or force. Another more intricate sense in which rape may be suggested by chaotech lies in the perceiver's reaction to what the machine used to be. Many people feel used, displaced, and oppressed by machines, whether that be the demands of the computer or the danger posed by the behemoths within the steel factory. The chaotech experience can be, although it need not be, one of reversal of that exploitation. A worker oppressed by a dangerous, loud, and dirty relationship to a machine may celebrate his eventual triumph over it as it lies in ruins. The chaotech experience may thus be one of perceived anger, violation, and triumph over machines that represent what was an oppressive way of life for some.

EROTIC: DEATH

If the chaotech machine embodies an order of desire, it is surely a desire for *death*. That is what has become of the machine. Its remains may have value, as do the shinbones of the saints for the faithful seeking relics, but it is only through a cataclysmic change that the chaotech object comes into perceptual existence. An inability to work any longer is death for a machine; it is the end of its functionality, the biggest change the machine will ever experience. If that state is aesthetically desirable, then death is the order of desire governing chaotech: death understood as The Big Change, as an alteration of status beyond repair, as crossing over to a new mode or dimension.

Del Caro's study of Nietzsche's "Dionysian aesthetic" identified it as an aesthetic of death, but of death keyed to change: "As the tragic god, Dionysus represents a heightened consciousness of transitoriness, and he is closely identified with death" (17–18). To the extent that machines reflect reading subjects, and that the decayed machine reflects a nostalgic identification, people may well be previewing their own deaths in chaotech. Baudrillard refers to the common literary device of "the robot that goes off the rails and destroys itself" as one in which "the spectacle of death is relished"—yet he notes the identification between such a ruined machine and the reader (*System*, 122).

Yet an erotic of death is paradoxical. Peckham explains the contradiction:

> The death wish is not a desire for death: it is a symbol of the desire for order on the part of men and women whose perception of the disparity between humanly created perceptual order and the demands of their transactions with the environment has resulted in an unbearable tension. The desire for death is merely the desire for the most perfect order we can imagine. (34)

This provocative statement can help us to understand the erotic of chaotech more deeply. The death of the machine brings it more fully within the control of the perceiving subject. We should remember that this may well be a subject who has experienced boredom, danger, or oppression at the hands of the machine. As anyone lost in the mysteries of a computer application knows, machines are very often *not* under the control of the user. And so a desire for the death of the machine is also paradoxically a desire for its complete transformation into a totally controllable form: simplified, passified, laid bare to the aesthetic reorganization of the subject.

Del Caro recognizes that same contradiction in Dionysian aesthetics, noting that "only where there are deaths, are there resurrections. . . . The reality of death is recognized, but its tyranny is not accepted" (23). The erotic of chaotech is then not only an anticipation of death but of the transformation that death may bring and the new thing that may arise from it (even if it retains memories of its past life). This aesthetic enables a rhetorical duality in which destruction and rebirth may both be supported by the same images.

MOTIVATING CONTEXT: RUST NEVER SLEEPS

Whether planned or not, obsolescence and then decay will come to every machine. This is true of every object and every creature in the world, of course. Such a realization is also consistent with Del Caro's explication of "a Dionysian world view, which emphasizes the beauty and necessity of change" (20).

In some ways one might see the process more clearly with machines, which people have made. Decay may be held off by constant rebuilding, refinishing, refurbishing, and so the person who is keeping an ancient Dodge Dart alive is vitally aware of the force of decay and what must be done to stop it. The decayed machine is a monument to *imperatives*: entropy *will* break down steel and copper and electromagnetic fields eventually. The motivating context suggested by chaotech is therefore one of imperative forces grinding on toward inevitable conclusions. As Kouwenhoven puts it:

> For both the arts and the sciences, in their different ways, seek to impose order upon the elements of experience our senses report to our minds. Insofar as they succeed, they temporarily confine elements that will ceaselessly batter and eventually demolish the design imposed upon them. The ultimate consequence of all design is therefore quite literally chaos—the violent eruption into randomness and asymmetry of energies that, for a time at least, have been forcibly restrained, even in the most open forms, by patterns of symbolic order. (231)

Kouwenhoven suggests that it is precisely because machines are made, because they are (unlike dandelions, which will also wither) intentional efforts to deny the imperative of decay, that their decay is so poignant and felt so keenly.

Many have felt, as does Joseph Tabbi, that "the image of the machine presents faceless and impersonal forces" (1). Those forces can be seen as powerful and sinister, as in Jacques Ellul's sweeping claim that "capitalism did not create our world; the machine did" (5). It is appropriate, then, that the decay of the machine also bears witness to the faceless and impersonal, and therefore irresistible and relentless, forces of decay in the world at large.

The urban scene can be constructed as a primary site where the imperatives of decay have overtaken machines and the society built upon them. Saper notes that urban decay is usually mythologized as *spreading*, imperatively and irresistibly (105). Narratives cast within that context tend, Saper notes, toward the tragic: "Wallowing in foolproof fate, the heroes of noir city know the *jouissance* in desolation, destabilization, and decay" (109). The "foolproof fate" of the characters is an imperative that parallels the inevitable fate of the city and its machines: they will all come to a grinding halt sooner or later.

Let me pause for a moment to summarize observations over the last few pages that suggest that chaotech may well be an *apocalyptic* aesthetic. In earlier work, I have identified some key characteristics of apocalyptic discourse, apocalyptic in the sense of recent discourses announcing the end (in one way or another) of the world. These discourses are very much keyed to a telic and determined view of history: the world *must* come to an end in this view, and although we in these last days may position ourselves to benefit from that apocalypse, there is nothing we can do to stop it (a kind of impotence, of course). Thus, both apocalyptic and chaotech share a strong sense of imperatives. Furthermore, no matter how sweeping the change and how terrible the final days, apocalyptic predicts a new heaven and a new earth, a changed world that will spring from the ruins. Thus, chaotech shares the apocalyptic view of death as engendering rebirth, of The End leading to A New Beginning. Chaotech may thus have roots in this very ancient form of discourse, predating the Age of Machines in which that aesthetic has flourished.

RHETORICAL POTENTIAL

As with the other types of machine aesthetic, some of the rhetorical applications enabled by chaotech have already become apparent. A decayed machine is, in the ordinary course of things, discarded and ignored. Yet chaotech preserves and resurrects these ruins. In that sense, chaotech may be the most intentionally rhetorical of the machine aesthetics, since the volition involved in preserving that which is destroyed can lead to its willed use in persuasion. We have noted before that chaotech depends especially on active participation by the perceiver to make the wreck into an aesthetic experience. Zucker reminds us of this in noting that

> the changing concept of the ruin is based not only on its objective appearance, but is equally dependent on the individuality of the beholder. His reaction will reflect his emotional attitudes, his cultural and intellectual level; but even more, the prevalent concepts of his time. (2)

This active cognition makes chaotech a particularly rhetorical act of image-making on the part of perceivers.

Machine as Alien

A clear and straightforward rhetorical use of chaotech is to present the machine as alien and horrible, a representation working against capitalism and the industrial culture it supports. Such images might speak especially to what Gregory Benford identifies as "a growing population that does not feel involved in technosociety" (225). Susan Fillin Yeh reminds us of the longstanding ambivalence towards the machine in the twentieth century (8–9). Certainly such ambivalence might be felt by factory workers who both live by and are enslaved by the machine. Chaotech exploits the negative side of that ambivalence, the feeling that technology and the machine is what is wrong with us. As Tabbi puts it, the feeling that "the most effective and potentially dangerous ideological force *at this moment* is to be found in those things we do every day to sustain the technological culture" (7).

We have noted that the identification between human and machine is especially strong in chaotech. Representing the machine as decayed and useless is therefore a way of rhetorically addressing the depravity and dysfunction of the human spirit. Tabbi describes how such images work: "The machine itself emerges as a metaphor, a figure representing forces and systems that the human mind and imagination cannot hope to master or comprehend, but for which we are nonetheless responsible" (20). If that metaphorical machine is a ruined one, the forces being represented are likely to be the dark side of humanity. One specific example of rhetorical uses of the decayed machine is *industrial music*, which Novotny identifies as part of "cyberpunk" culture (which we consider below) (114). This industrial music goes right to a sense of the machine as alien, for it uses mechanical sounds in ways specifically designed *not* to be cozy and lilting, to project an image of a machine off the track and flailing itself to pieces. Novotny explains that

> industrial music is impressed with parody, pessimism, and detachment, providing the dystopian background music with which to survey the postmodern cultural landscape . . . industrial music is situated in the context of postmodern urban detritus and decay. . . . Industrial music's brittle and mechanized sounds of detachment reflect the brutality and despair of dystopian, postmodern life. (115)

In contrast to the 1960s cheery use of engine sounds in Jan and Dean drag racing songs, the machines of industrial music are violent and alien.

Against the Urban

Rejection of the city might be a rhetorical goal for many reasons; one may want to paint the city as bleak and failing so as to reject the industrial-capitalistic economy, in reaction to the racism and classism prevalent in the city, and so forth. As Saper notes, images of urban decay inherently serve political, rhetorical purposes: "For liberals, the urban decay causes the social problems; for racists, the people cause the urban decay; and for neoconservatives, liberal policies cause the urban blight" (96). For instance, Frances Bonner notes the racist rhetoric signalled by chaotech in two popular films: "Ethnic heterogeneity is used as a sign of the decayed world in *Blade Runner* and *The Running Man*" (199). The decayed machine can thus become a symbol for the decayed city as well as for the decayed social and economic relationships that the city represents.

Saper identifies the literary and filmic genre of *noir city* as one that makes heavy use of images of decay and collapse. Saper argues that

> one could make a list of noir city's ingredients: bittersweet irony, gritty realism with a wink, simple assertions spiked with startling metaphors, obsessive planning and rationality falling prey to fate's unlucky contingencies, and dark ominous urban settings peppered with suspicion and seduction. (110)

Clearly, chaotech contributes to most of those "ingredients": it is an ironic aesthetic, it cultivates gritty realism, it lends itself to metaphorizing machines, it celebrates the collapse of mechanism before fate's contingencies, and above all it is part of dark and ominous urban settings. The city is a focal point for those arguing that the train of technological and industrial progress has ended. In urban contexts especially, decayed machines may be used to signal that, as Siegfried Giedion claims, "faith in progress lies on the scrap heap" (715).

Recovering the Human

Another rhetorical application of chaotech is to recover an emphasis on the human in social and public discourse. This use is predicated upon the idea that society needs a balance between mechanism and human needs, yet in the Industrial Age the balance has shifted toward the former (Giedion, 714–24). Turkle points to "widespread fears about machines and the dangers of too intimate relationships with them" (*Second*, 205). People may fear that machines are better than they, and need a redress of that disempowerment. As Joe Sanders notes, the machine "represents an emotional challenge to people nearby. We wonder just how many things the machine can do that we cannot" (168).

During the heyday of mechtech, Lewis Mumford observed a "monstrous dehumanization" caused by factory machines (*Technics*, 145–46). As Stuart Ewen notes for our more electrotech world, "The automated workplace has, for many who must march to the digital beat, led to a failing sense of self" (187). If, as L. Casey Larijani notes, "To many people dimming the line between virtuality and reality is scary" (187), chaotech may be used to speak against the electronic machines that create virtual reality. Novotny points to a more physical, and to many disturbing, overwhelming of the human by the electrotech machine in terms of "interfaces of human and computer technologies through prosthetic limbs, implanted circuitry and genetic alterations" (107). As these trends continue people may well fear "humanization of machines," in which as Sanders notes, "different levels and varieties of humanness . . . can be projected into machines" (167). Science fiction writer Brian Aldiss discusses the ways in which robots are imagined as sinister human-machine amalgams in popular literature and film, another indication of concern over a loss of humanity to the forces of technology (3–9). Ullman expresses the realization that human-machine meldings do not allow the full expression of the human spirit: "But the computer is not really like us. It is a projection of a very slim part of ourselves: that portion devoted to logic, order, rule, and clarity" (89). Depiction of machines in decay evens the score; it points an audience toward the other, more vital half of that equation by representing machines as aesthetically repulsive.

Cyberpunk

The rhetorical applications of chaotech to issues of alien machines, urban decay, and a loss of humanity all come together in a recent genre across several arts known as *cyberpunk*. Mike Featherstone and Roger Burrows explain that "the term cyberpunk refers to the body of fiction built around the work of William Gibson and other writers, who have constructed visions of . . . the dark side of the technological-fix visions of the future" (3).

Cyberpunk warns its audience of coming *dystopias*, in contrast to more traditional science fiction, as Novotny explains: "Cyberpunk is a radical disjuncture with traditional science fiction imagery, projecting dystopian images of decrepit and corrupted technology" (107). It warns its audience of a bleak future if trends continue; as Fredric Jameson explains the wider genre, " . . . the dystopia as a form is better understood in classical science-fiction terms as the 'if this goes on' narrative, which extrapolates a certain feature or tendency in contemporary society and then, as in an experimental laboratory, fast fowards its development" (37). Dystopian visions, cyberpunk among them, thus have in Jameson's view the function of "articulating a social structure in full evolution" (38)—and not a positive evolution. In other words, cyberpunk is a commentary on our own problems and dysfunctions. As Mark Poster observes, cyberpunk reacts "against depictions of cyberspace as utopia: the wounds of modernity are borne with us when we enter this new arena and in some cases are even exacerbated" (86).

Many characteristics of cyberpunk might be identified. Hafner and Markoff explain that "in cyberpunk novels high-tech rebels live in a dystopian future, a world dominated by technology and beset by urban decay and overpopulation" (9). Lewis Shimer lists "this rock-and-roll quality—the young, hip protagonists, the countercultural attitude (symbolized by the ever-present mirrorshades), the musical references" (21). Woolmark finds similar characteristics in "the hackers, street wise rock'n'roll heroes who wear mirrorshades and do 'biz' in the urban sprawl, dealing in designer drugs, information technology and stolen data" (111).

We have noted how chaotech reflects a postmodern sensibility. The style of cyberpunk writing does the same: "Cyberpunk science fiction is the literary incarnation of postmodernism's eclecticism and decentering, reflecting the shifting contours and disintegration of postmodern culture" (Novotny, 102). As Woolmark explains, "The multiplicity of signs in cyberpunk emphasises surface style as an emblem of a depthless present, and as such it is taken as an indication of the instability of language itself" (108), a hallmark of postmodernity.

Machine aesthetic plays a major role in cyberpunk, for as Bonner points out, nature is almost totally absent in much cyberpunk, replaced by machines and the cityscape (199). Novotny explains at length the ways in which cyberpunk uses chaotech as a reply to and refusal of a high-tech corporate culture:

> Cyberpunk's main themes focus on the forces of cultural integration, such as media and communications technologies, massive computer and satellite networks, and multinational corporations. Contrasted with the images of immensely powerful multinational corporations and vast computerized networks are the street worlds of the punk underground and urban subcultures. Cyberpunk's settings are reminiscent of the amorphous and decaying urban wasteland of the postindustrial and deindustrialized pres-

ent, with pervasive images of imminent collapse. Its imaginative themes reflect the dystopian context of postindustrial, technological culture. (103)

Chaotech images are clearly found in Novotny's depiction of cyberpunk's "industrial landscape of empty warehouses and abandoned industrial plants, of dead grass in the cracks of broken freeway concrete alongside the broken slag and rusting shells of refineries" (106), a cityscape of "dinginess, squalor and filth" (114).

The troubling merger between humanity and the machine is a major theme in cyberpunk. Woolmark finds that "the narratives of cyberpunk . . . seem to reveal a deep anxiety about the disintegration of the self" (114). Featherstone and Burrows note that in cyberpunk,

> the key analytical categories we have long used to structure our world, which derive from the fundamental division between technology and nature, are in danger of dissolving; the categories of the biological, the technological, the natural, the artificial *and* the human—are now beginning to blur. (3)

For this reason, David Tomas notes the tendency of cyborgs in literature "to function in hostile, dystopic, futuristic worlds governed by various kinds of renegade military/industrial or corporate activity" (21). Many of those sinister cyborgs are depicted as puppets of corrupt and repressive regimes (38–39).

Perhaps the best known cyberpunk artist is the novelist William Gibson, author of the classic *Neuromancer*, among other novels and screenplays. Featherstone and Burrows also locate his work, and thus cyberpunk, within the postmodern context: "The work of Gibson has been held up as the prime exemplar of postmodern poetics" (7). Nigel Clark describes Gibson's work in clearly postmodern terms as "a sea of shifting, flickering signifiers, they defy a stable reading, and must be decoded from moment to moment" (122). Featherstone and Burrows report that Gibson has been hugely influential, his ideas being read as social theory, his works familiar within the Pentagon, his novels studied by business leaders (8–9). Key to Gibson's postmodern style is his vision of a world in which signs are paramount. As John Christie explains,

> The stylistic complement to this postmodernity was a postmodernism—an artful, conscious combination of surface description, multimedia intertextuality, autoreferentiality, and so forth. The

discernment not just of an age when image and appearance took over reality, but when the simulacrum ... invaded and subverted inherent notions of identity, history, all relational coherence. (173)

As we discovered in the chapter on electrotech, a takeover by the simulacrum is effected by electronic technology. Thus, Gibson's work is a commentary on machine aesthetics, and it often uses the aesthetic of chaotech as a response to and refusal of more conventional aesthetics.

Others have noted the juxtaposition of high technology with the collapse and decay of the city in Gibson's work. Clark observes, "Recalling the set design of the film *Blade Runner*, *Neuromancer* and its sequels feature a series of hyper-aestheticized urban spaces: terrains in which vibrant new signifying surfaces are layered over the detritus of obsolescent forms" (121). Carol McGuirk claims that "Gibson's cyberpunk fiction [critiques technology's] power to gratify human desires" through "technological interventions that palliate or conceal some perceived or real defect in the self or in the soul" (113). That critique is very often carried out through bleak chaotech images of the decayed city. As Novotny observes, in Gibson's work "the images of urban decay and disintegration reinforce the sense of social fragmentation and bleakness which has been central to cyberpunk" (105). Gibson's work, in sum, illustrates quite well the rhetorical potential for a critique of society and technology inherent within chaotech aesthetics.

CONCLUSION

We have seen that chaotech is a machine aesthetic that plays off of other aesthetics of mechtech and electrotech. This ironic aesthetic must be experienced in terms of what the machine was and what it used to do. It may be clear from our review of the dimensions of machine aesthetics here that chaotech often serves the rhetorical purpose of refusal, rejection, or rebellion. Considering the ways in which mechtech and electrotech have been allied with empowered and established interests in society, with dreams of progress and utopia, with power and efficiency, envisioning the decay of mechtech and electrotech tends to serve more revolutionary and radical interests.

Given chaotech's negative connections to mechtech and electrotech, one might expect to find its images often commingled with our first two aesthetics. Indeed, machine aesthetics may occur quite often in mixed modes. It is important to be clear theoretically about the different ways in which mechtech, electrotech, or chaotech manifest the dimensions of machine aesthetics. But a particular experience or text is likely to involve more than one of these aesthetics. Sometimes the ways in which the aesthetics interact may be important in influencing the rhetorical uses to which they are put.

The critical application to which we now turn is an especially rich example of all three machine aesthetics working together in mixed modes to create some powerful rhetorical effects. The 1985 film *Brazil* (Gilliam) creates a distinct aesthetic within its text that is based largely on mechtech, electrotech, and chaotech. We will turn to that as an example of the theoretical concepts developed so far.

As we turn to an application of machine aesthetics in actual texts, it may be useful to consider more generally how machine aesthetics emerge in texts, with an eye to how the critic might apply the theory developed thus far. The central idea I would defend here is that machine aesthetics may be experienced in direct experience, in representations of experience, or in simulational experiences. Direct experience, of course, is seeing, touching, using an actual machine or object that partakes of machine aesthetics. Representations of machines, such as photographs of huge industrial engines or locomotives, the latest camera or laser pointer, selectively highlight certain dimensions over others. Such representations have a venerable and ongoing role in advertising, for instance, as they represent to the public the shiny steel flow of the latest sports car's skin. A simulational experience is, as Jean Baudrillard explains in his classic book *Simulations*, without reference to any reality beyond itself. A session in a virtual reality machine would be a clear example, although Baudrillard points out that increasingly our culture, politics, and everyday practices are so self-referential, so exclusively textual, so mediated and aestheticized that they must all be regarded as simulations. As I will argue in the next chapter, *Brazil* must be regarded as just such a simulation for the most part.

Any of these three kinds of machine aesthetic experiences may serve rhetorical purposes. Stalin built huge hydroelectric dams for the purpose of impressing his own citizens and the world with the industrial might of the Soviet Union. An illustrated advertisement may persuade consumers to buy a computer by stressing its ability to serve the impulses of the will instantaneously. A film might warn its audience against the horrors of nuclear war that exists nowhere in reality by depicting a devastated chaotech world in the aftermath of a holocaust. In all these cases, a critical analysis of the experience, representation, or simulation would profit from considering how machine aesthetics served rhetorical purposes.

I propose here no dogmatic or limited critical method. My purpose so far has been to develop a language for the critical understanding of the rhetoric of machine aesthetics, to open up for the critic an understanding of what the experience of machine aesthetics might mean. But that understanding can be applied in many different critical methods, or in no specific method at all beyond a close reading of a given text. Awareness of machine aesthetics is offered here, then, as a tool in the critical workshop, to be used in conjunction with other tools and with the themes and meanings immanent within a text that the critic wishes to explain. I therefore take no stance on whether a given machine aesthetic must be intended or not, culture-specific or not, expressed verbally or depicted in images, and so forth; those are all issues that will vary from one critical practice to another. We turn now to an analysis of *Brazil*, showing how awareness of machine aesthetics can help the critic to see meanings and effects offered by the text.

CHAPTER FIVE

Simulations and Machine Aesthetics in *Brazil*

The 1985 film *Brazil* (Gilliam) is a remarkable work. Set "somewhere in the twentieth century," it creates an Orwellian society of crushing fascism. Big business and big government have merged. Nearly everyone works for one monolithic agency or another. This society worships bureaucracy and hence is drowning in paper, proper procedures, policy, and office work. The world has been paved over and built up; urban sprawl consumes everything. Conversation, entertainment, indeed all of everyday life is swallowed up in trivia, parties, plastic surgery, obsessive television watching, and paper-pushing. Terrorist bombings break into this dystopia from time to time, and the threat of these attacks anchors the society's ideology: individual rights are gone, police tactics reign supreme, in a state wholly given over to combatting "terrorists." The film follows Sam Lowry, petty bureaucrat, over the course of a few days as he goes from office drone to wanted criminal to torture victim.

The film begins with a fly dropping into the works of a machine that is typing out names of the latest citizens to be rounded up by the Department of Information Retrieval. Information Retrieval is located within the Ministry of Information, the dominant controlling force of this society. The agency specializes in capturing and torturing thousands of ordinary citizens in the paranoid hope that information about terrorism may be gained. Lowry's friend Jack Lint works there as a torturer. The fly in the typewriter causes the machine to change the name of one suspect, Archibald Tuttle, to Archibald Buttle. The blameless Buttle is arrested while police storm his apartment. His neighbor, Jill Layton, witnesses the whole affair and commits herself to the wholly frustrating task of filling out forms and going from one agency to another in an attempt to gain Buttle's release. This effort soon earns her the attention of Information Retrieval.

In this bleak society minor bureaucrat Sam Lowry (Department of Records, also within the Ministry of Information) breaks the monotony of his day with fantasy dreams about flying (literally) to the rescue of a maiden trapped by monsters. He has no social life beyond visiting his wealthy mother, whose calling in life is shared by her rich friends: going from one party, restaurant, and plastic surgery to another. One day, Lowry takes some forms to the widow of Archibald Buttle for her signature and there he sees a woman who looks exactly like the damsel in distress of his dreams: Jill Layton. However, when he tries to talk to her, she flees from him because of his identification with the bureaucracy. In an effort to meet her, Lowry reluctantly accepts a transfer and promotion from Records to Information Retrieval, which would allow him better computer access to information about Layton.

Lowry uses his new position to locate information about his dream woman. He discovers that she is wanted by Information Retrieval because she questioned the false arrest of her neighbor, Archibald Buttle. Lowry meets the woman, Jill Layton, but she rejects him once again. Lowry's pursuit of her, fueled by passion rather than proper bureaucratic form, begins to violate one law, one office policy, after another until he himself is under investigation by Information Retrieval. To add to his woes, he allows a renegade heating repairman to fix a defective thermostat in his apartment. This transgression is discovered by Central Services, and becomes further proof of Lowry's terrorist intentions, since the repairman is none other than Archibald Tuttle, being sought by Information Retrieval.

Attempting to alter computer records to clear his and Layton's name, Lowry is nevertheless captured and Layton is killed while they hide out in his mother's apartment. Layton is taken to the torture chamber where Jack unwillingly at first, but then with proper professional dispatch, proceeds to do his work. The shock is too much; Lowry loses his mind at the end of the film and we leave him lost in a permanent fantasy world.

It is clear that *Brazil* is a cautionary tale against fascism and the police state. The film is *about* any society in which big business and big government become *too* big, and therefore it is about *us*, the audience living in late twentieth century capitalist societies. The alliance between government and business is clear. Television ads offer new ductwork from Central Services, which is both government agency and a purveyor of heating and cooling systems. The film takes place around Christmas time, and it is clear that this commercial holiday is now serving the interests of the state as well as commerce. The Salvation Army, marching by playing Christmas carols, has been renamed "Consumers for Christ" according to their banner. Torture victims are referred to in commercial terms: "The next customer has been delivered," Jack Lint's secretary tells him. And as Sam Lowry himself is strapped into the chair, a guard tells him, "Don't fight it, son, confess *quickly*! If you hold out too long you could jeopardize your credit rating."

It is also clear that *Brazil* is relevant to many industrial societies. Signs within the text reference a wide range of times and places within the last fifty years. A caption in the first few seconds identifies the story as taking place "somewhere in the twentieth century." The accents are all British. The clothing and grooming are all 1930s and 1940s American, with wide lapels, floppy brimmed men's fedoras, stylish women's hats and snoods. The military/police forces are either late twentieth century black suited, black helicoptered, M16-ed American, or else in their comic opera uniforms and high brimmed hats they reference the 1930s and 1940s (the heyday of unembarrassed fascism) of Hitler and Mussolini. The predominance of computers bespeaks our time, of course. Just to keep the audience guessing, there is the title, *Brazil*, which for many connotes a distant, vaguely familiar land, probably run by generals, that one is sure one has heard something about over the years. And the swamp of bureaucracy and paperwork that is this society could reference many nations from Sweden to the UK; even as poor Mr. Buttle is taken prisoner, a bureaucrat thrusts forms at the frightened Mrs. Buttle that must be filled out in conjunction with his capture. This multivalent relevance seems intentional; the film speaks to any and all industrial societies facing dilemmas of growing industrial and military power in the last twentieth century.

Brazil seems especially to speak to societies that share its preoccupation with terrorism. An interview with the Deputy Minister of Information early in the film indicates that seven percent of the Gross Domestic Product of this land is expended in the pursuit of terrorists. The citizens explain away their lack of liberties and the drabness of their lives with claims that these privations are necessary to combat terrorism. Jill castigates Sam for working in the Office of Information Retrieval; "I suppose you'd rather have terrorists," he replies huffily. To which she returns, "How many terrorists have you met, Sam?" One might easily think that *Brazil*'s obsession with terrorism as a policy justification is very close to our own.

In sum, it is clear that this film warns its audience of the pitfalls and seductions of political and economic fascism that might befall them. What may not be so clear are the rhetorical mechanisms that advance that thesis. I propose to show in this chapter that *Brazil* tells us that fascism is propped up by a culture lost in *simulations*. Furthermore, in *Brazil* (and by extension in our own societies) those simulations are based on a machine aesthetic. My purpose will not be merely to identify machine aesthetics in the images and events of *Brazil*, and certainly will not be to look for machine aesthetics merely for the sake of illustrating the theory developed in earlier chapters. Nor will I pursue a cookie cutter approach of trying to identify every dimension of each type of machine aesthetic. Instead, I want to use a critical awareness of machine aesthetics to describe the warnings the film gives us about how simulations contribute to fascism. Although machine aesthetic need not support repression and conformity, we saw in earlier chapters that it contains the potential for such a rhetorical application. *Brazil* thus exemplifies the dark side of machine aesthetic.

In general, mechtech and electrotech aesthetics are signs of power in *Brazil*, and chaotech is evidence of the ultimate failure, rottenness, and corruption of that power. But the people of *Brazil* are not able to see beyond those aesthetics; they are by and large trapped within a simulational culture supported by machine aesthetics. I will suggest toward the end of this chapter that the film not only warns its audience against such infatuation with simulations, but does so for an audience that is experiencing a sort of demi-simulation itself, namely *Brazil!* We turn now to think about the simulational nature of *Brazil*'s culture, and perhaps our own.

SIMULATIONS

As Jean Baudrillard explains in his book *Simulations*, a simulation is an experience that is largely self-referential, with little external meaning. The question "What is this about?" is irrelevant in a simulation. As we noted in chapter four, simulations may usefully be contrasted with direct experience and representations. A direct experience is just that, such as stubbing one's toe. A picture or story about stubbing a toe is a representation of such an experience, but as a representation it is nevertheless about something and has meaning beyond itself. A simulation is another kind of direct experience, but an experience that is mediated, constructed, artificial—in short, an exclusively *aesthetic* experience. It is stubbing one's toe in a computer game as one (in the virtual body of, say, a wizard) chases a monster through a dungeon. Neither wizard, nor monster, nor dungeon refers to anything real. The point is to enter into the simulation itself.

As Baudrillard and others have noted, late industrial capitalist cultures may be accused of becoming increasingly simulational. Our politics seem not to be about the distribution of real goods and services but rather about who is winning or losing within the hermetic circle of politics itself. Political struggles are covered in terms of the personalities and strategies of those within the struggle as or more often than in terms of the real issues to which they might connect. Presidential debates are followed not for the real insights they give about actual issues—they give none—but for how each candidate presents himself within the debate itself: Is the clothing correct, the grooming tidy, and are gaffes avoided? Increasingly, politics is about itself, and is thus simulational.

A culture obsessed with entertainment is simulational, as is ours. The phenomenon of "being famous for being famous" is well known and simulational, for it has no reference to achievements or merits beyond the screen. Children immerse themselves for hours in fortresses, castles, haunted islands, and battlefields that exist nowhere but within computer simulations. Daytime television viewers grieve over the deaths of friends who have never existed anywhere but in a nineteen inch screen. We become lost in sporting spectacles that do not reference anything outside themselves; what is the Super Bowl about if not the Super Bowl?

Above all, we lead *aestheticized* lives: We dress like a cowboy today, not to reference real cowboys or any actual experience we have had on the range but simply for the sake of the aesthetic experience itself. We decorate our homes like Hawaiian plantations which we have never visited, and will change all that tomorrow by moving in furniture and paintings from an Art Deco period we have never experienced. Our concerns, this critique holds, are no longer with what is right or wrong or true or false but with what is entertaining, pleasing, aesthetically appealing.

It is clear that the world of *Brazil* is highly simulational. Simulations depend on mediation to create worlds electronically that the audience may enter. For *Brazil*'s audience as well as characters, an important instrument of simulation is the television. Television sets are *everywhere* in *Brazil*, and they seem to be constantly on; perhaps they cannot be turned off. These sets purvey nonstop simulations to people, largely in the form of old movies and film shorts (The Three Stooges, *Casablanca*, and so forth). The very first images of the film are of the cloudy world that is Sam Lowry's dream fantasy.

The particular set we are watching near the start of the film goes from a commercial to an interview with the Deputy Minister of Information, who is talking about the latest terrorist bombing (including one in the very store in which the set is located). But it becomes clear that although terrorism is taken seriously, it is also taken as a simulation. For this society, both terrorism and the fight against it are a game. The point of engaging in or fighting terrorism is simply to do so for the sport rather than for any external political meanings or the hope of social change. The Deputy Minister castigates the latest bombing as "bad sportsmanship. If these people would just play the game they'd get more out of life," and he goes on to describe the efforts against terrorism using sports metaphors.

Ubiquitous television sets are an ongoing sign of simulations in the film. Lowry at first experiences Jill Layton more often in such mediations, as an image, than he does first hand. He first sees her, the woman of his dreams, in a television monitor as she enters the Ministry of Information. He next will see her reflected in a piece of broken glass as he sits on the floor of the Buttle apartment and she peers down into it from a hole in the ceiling above. The porters in the Ministry of Information are likewise immersed in televised simulations. These sinister characters sit high above the humble supplicants who come to them (hopelessly) for information, and all peer into television sets that show them the very person who is in front of them.

Within *Brazil*, sheer escapism into mediated, aestheticized simulations rules the day. At the Department of Records, the chief occupation of the thousands of bureaucrats is to watch old movies on television. Only when the boss, Mr. Kurtzmann, peers out of his office door is there an instant reversion to shuttling piles of paper from one place to another.

The people of *Brazil* willingly embrace simulations in preference to the dreary reality they keep at bay. Lowry joins his mother at an expensive restaurant complete with French waiters. The food is advertised as the usual classy fare: steak, duck a l'orange, and so forth. The food arrives with much fanfare on silver servers. But it turns out to be little round dollops of mush, each colored differently, with a photograph of what they are supposed to be beside them on each plate. The diners have so embraced this simulation that they don't even notice what is before them: Lowry, having ordered a rare steak, pokes at his mush balls and complains, "It isn't rare!" Indeed, it isn't even *steak*, but the diners are lost in the simulation and cannot see the difference.

The hyperindustrialization of this society has destroyed the natural environment, but simulations overcome that difficulty: Vehicles travel on elevated highways over polluted slag heaps that they never see, for the roadways are completely lined with colorful, cheerful billboards. These billboards create an artificial, aestheticized roadway that shields drivers from the real mess beyond. This society also has its poor, but simulations are offered for their benefit as well: The Buttles live in a miserable, squalid apartment complex grandly named "Shangri-La Towers."

A preoccupation with bureaucracy is a powerful kind of simulation. One becomes lost in the language and rationale of paperwork and forms, losing contact with direct experience. Mr. Kurtzmann is describing the bureaucratic status of the unfortunate Buttle to Lowry: "The Population Census has got him down as dormented, the Central Collective Storehouse computer's got him down as deleted, Information Retrieval has got him down as inoperative, Security's got him down as excised, Administration's got him down as completed—." "He's dead," Lowry cries at this point. "Dead?!" replies Kurtzmann, incomprehendingly, who cannot handle the brute fact in place of bureaucratese. Kurtzmann prefers to spin his wheels in the comforting simulations of policy and procedure.

Lowry himself spends a lot of time in simulations within simulations. He has a recurring dream fantasy, to which he repairs both night and day, in which he is a winged warrior flying to the rescue of the beautiful Layton, trapped inside an iron cage. This simulation is all powerful and likely to intrude upon his other (simulational) experiences at any moment. When Lowry confronts a soldier who is about to take Layton away under arrest, the soldier suddenly turns into a monster from his fantasy, which Lowry then battles. Earlier, Layton calms Lowry's fears that the police are "lying in wait" for them by exclaiming, "Christ, you're paranoid. You've got no sense of reality." Indeed he has not, nor have most of his fellow citizens. Nor has Lowry any desire to ground himself in that reality. He tells Layton, "We should take the lorry and get away." "Where to?" she replies. "Anywhere as fast as possible," says Lowry. "There isn't anywhere," she says, a telling response that indicates the hermetic nature of the simulation. "Just far away," returns the mournful Lowry, to which she says, "That's not far enough." Indeed it is not, for wherever "it" is will still be within the simulational culture.

One final point is needed to comprehend this totally simulational culture. One of the major ideological tenets of the society of *Brazil* is a *machine metaphor*. Several references within the film make it clear that the characters regard their society as if it were a machine.

Ongoing events and expressions in the film clearly say that this society is a machine. The claim is made more or less explicitly a couple of times, when Lowry's dead father, an influential bureaucrat, is fondly remembered by a coworker as "the ghost in the machine." But the best reference to the metaphor is, I think, the ongoing signs and exhortations to conform and cooperate. A machine, whether mechtech or electrotech, requires its parts to work together for the common good. In *Brazil*, the slogan "We're all in this together" is found repeatedly on billboards, and is even used by the renegade repairman Tuttle twice when speaking with Lowry. Conformity is expected even down to clothing: staffers in the Department of Information Retrieval all wear the same distinctive suit, as they do in Records and in every other bureau. Subordinates take their cues from superiors: when Mr. Warrenn, Lowry's new boss in Information Retrieval, puts his jacket over his shoulder, all his staffers do the same.

There are other reminders of the machine metaphor in this society. One does not call repairmen for difficulties with plumbing or ventilation, one calls an *engineer*. Despite the fact that the society is floundering in bureaucracy, the call for efficiency is nevertheless repeated often, as in Mr. Warrenn's exasperated demand of Lowry, who is swamped in paper, "What the hell is this mess? An empty desk is an efficient desk."

An important mark of any machine, mechtech or electrotech, is quantification: machines are measured, they are given instructions mathematically, and their outputs are calibrated. This society is enamored with enumeration: as many things as possible are expressed numerically, which is consistent with a mechanical model. In the fancy restaurant with his mother, Lowry is told by the waiter that one *must* order by the number on the menu. When Lowry refuses it completely undoes the waiter, who ends up hissing the number himself. Bureaucrats are known by their badge numbers more than their names. Even the most ordinary occurrence requires enumeration. Lowry stops Central Services workers from entering his apartment by asking, "Have you got a twenty-seven B stroke six? I'm a bit of a stickler for paperwork. I mean, where would we all be if we didn't stick to the correct procedures," which is a sentiment appropriate for operating machines.

To sum up so far: *Brazil* depicts a fascist society locked in the inescapable fantasies of simulation. An important part of that simulation is the metaphor of society as a machine. This way of seeing that society enables us to understand one of the major underpinnings of this particular kind of simulation: machine aesthetics. We turn now to consider the widespread evidence of mechtech, electrotech, and chaotech aesthetics in the film. We will examine images of these machine aesthetics so as to maintain this string of connections among several themes: *Brazil* argues that fascism depends not only on simulation, but on a simulation based on machine aesthetics.

FASCISM, SIMULATIONS, AND MECHTECH

Mechtech images abound in *Brazil*, creating an ongoing aesthetic theme that supports the fascist society. Government buildings and private apartment houses alike (there seem to be no freestanding homes) are made in the same plain, boxy, geometric lines in neutral machine-like colors of black and gray beloved of mechtech architecture. Except for Lowry's fantasies, we see nothing but urban and industrial cityscapes; pastoral countrysides are long gone. Buildings for agencies and for the rich and poor alike are huge, massive, signifying strength and solidity. This architecture lends meanings of strength and stability to the state. Everything is on a communal, collective scale; the individual shrinks and becomes a part of the more important whole.

This society is a highly mechanized one; tools, machines, and automatons are everywhere. Sam Lowry's apartment is highly mechanized. As he awakes his clothes trundle out of the closet on a mechanical rod, coffee and toast are being prepared automatically, and the shower turns itself on as he enters. Archibald (Harry) Tuttle, the renegade repairman, begins most jobs by taking out an enormous motorized screwdriver with a flourish. Mr. Helpmann, the Deputy Minister of Information, goes around in a wheelchair. And on a humorous note, a simple machine drops a weight on a triangle, one side of which is marked "Yes" and the other "No." It appears in several places, including Lowry's new desk in Information Retrieval and on the porter's desk in the lobby, where he plays with it while placidly frustrating the supplicants who come to him for help in navigating the bureaucracy.

Mechtech machines are marked as integral to this society's life by images that make machinery integral with the physical body. The secretary in Jack Lint's office is a friendly, motherly sort who smiles sweetly at visitors while she transcribes the screams and pleadings of torture victims, conveyed to her from the torture chamber through headphones. But the hands with which she is rhythmically, mechanically typing are enclosed in a steel structure of splints and wires—she is becoming a machine as she records the horrors next door. Lowry's mother keeps trying to interest him in Shirley, a friend's daughter. Shirley has an enormous metal external dental brace surrounding her face. Even Lowry's dream fantasy encases him in wings that depend on mechtech mechanisms of metal and gears. The ubiquity of machines, even to the point of merging them with the human body, expresses the importance of clockwork order, of precise, planned, and efficient operation, that is essential to a fascist state.

The mechtech machines of *Brazil* are designed so as to feature the dimensionality of surface and depth. We can see the inner workings of many machines. The film is a feast of small visual detail: springs, wires, levers, sturdy webs of steel and wire are visible everywhere within the ubiquitous machines. The shower in Lowry's apartment does not merely turn on automatically, but we see the gears and springs that work the hot and cold water handles. Ductwork is everywhere, in offices and private homes alike; they have come out from their usual hiding places in walls and are plainly visible, giving a texture to the visual background of nearly every scene. The metaphor of society as machine is thus supported by the presence of machine parts in every scene of everyday living.

The fascist society of *Brazil* depends on an aesthetic that means power, permanence, impregnability, and the threat of violence. Those meanings are transferred to the government-industrial complex, which keeps citizens under control through those aesthetic means. Mechtech dimensionality shows the strongest inner workings of the machine in those places where the machine of state is the strongest and most sinister. The basement of the Ministry of Information houses the SS-dressed secret police units, done up in military black or in their Pinochet cum Hitler dress uniforms. Here we see massive steel and iron gears, wheels, rivets and bolts holding it all together, pistons, boilers, girders, huge coiled springs—the foundations of the Ministry house its source of violent power, the police, but that scene has also become a mechtech machine itself, and we see its inner workings.

State violence is consistently associated with mechtech aesthetics in the film. Guns appear frequently, large black mechtech devices used to frighten and, if necessary, kill the cowering public: the police, of course, have them all the time, and Harry Tuttle pulls out his Luger quite unnecessarily as he forces his way into Lowry's apartment for the first time to insist upon making a repair. Violence, specifically male violence, is never far away in the film. Helmeted, black-clothed soldiers burst frighteningly into Buttle's apartment early in the film, and into Lowry's mother's apartment to take him away towards the end.

Mechtech is a male aesthetic, and the society is dominated by males and male violence (in one scene a one-legged woman must stand in a subway car while surrounded by seated, uncaring men). Prisoners are trussed up into steel and canvas bags, circled with chains and straps, hoods with breathing tubes are tightened over them and they are hung on hooks for transport; it is a sort of mechtech straitjacket. The torture chamber itself is a mechtech shrine: it is an enormous metal egg. There is no floor, only a web of pipes. A narrow steel bridge brings one to the torture chair, a wicked contraption of metal with wheels and gears all around it, straps and a steel cage for restraining the body and head. On a tray by the chair lie mechtech instruments of pain: a drill, pliers, probes, all metal, all shining bright, all deadly. Not suprisingly, the populace goes in terror of this state violence; during several episodes when it appears as if he might be questioned by the police, Lowry nearly panics in fear. Thus, fascism is propped up not only by violence but by a violent aesthetic.

When Lowry helps Layton escape from sure arrest by the police for the first time, they make their getaway in a huge, massive truck driven by her (it seems this is her job). The truck (or lorry) is not hers but owned by the state; its size, sheets and blocks of steel, and mechtech pipes, knobs, wheels, and other appurtenances make it a moving symbol of enormous mechtech power. In this truck they drive to a nearby power plant, where Layton is to pick up a package. Lowry is terrified, associating the power plant with the police: "They'll be lying in wait!" he cries. And when they get there, this place of state-sponsored fear is practically an amusement park of mechtech: miles of pipes, tall metal towers circled round with tubes, pipes, gauges, and gears, vents and doors from which gas and fire is expelled rhythmically, other massive vehicles and machines working frantically. It is a place of fear created by an aesthetic of machine, and thus of state, power.

We may recall some of the marks of mechtech subjectivity and production. Those aesthetic markers are present throughout the film, conveying the message that the society is a machine and the people within it either working parts or uniform product units. Two scenes are especially suggestive of the rhythmic, choreographed movement of a working machine. When Mr. Kurtzmann, the supervisor for the Office of Records, is watching, his minions scurry busily around the office, moving piles of paper here and there. Their movements are precise and rhythmic. They are practically dancing, but they also suggest the spinning, articulated parts of a well oiled engine. A machine-like chattering keeps time to their movements. In another scene, Lowry has arrived in the bleak hallway of the thirtieth floor of Information Retrieval. He stands in the barren hallway, wondering what he is to do. In the distance he hears a faint commotion, and sees a block of moving men pass swiftly along corridors just glimpsed at a distance. Soon this procession comes down his own hallway. It is Mr. Warrenn his new supervisor, strutting purposefully in front of about a dozen assistants. The assistants are rhythmically waving paper and shouting questions at him; he is just as rhythmically barking out, "Yes! No! Tell him I want a receipt!" in reply. They seem like a locomotive made of humans; Lowry is caught up in this meat mechtech and delivered to his own office.

The aesthetic dimension of specialized, precise function is found in another episode. Lowry goes to lunch with his mother, and the restaurant is bombed. His fellow diners ask him if he can't "do something about those terrorists," but he replies, "It's my lunch hour; besides, it's not my department," thus displaying the kind of specialization and singularity of function that is a hallmark of mechtech machines and machine parts. Plainly, the values of conformity and cooperation in this society are reinforced by an aesthetic that enlists people to engage in machine-like behavior.

Uniformity of produced copies is a hallmark of mechtech aesthetics. That uniformity is used in *Brazil* to encourage the kind of conformity that props up fascist states. Uniforms have a prominent role in the aesthetics of this government, of course, with the predominance of police and military units. The soldiers are faceless copies of one another, black helmets completely obliterating any distinguishing features. But even the clothing worn by civilians mark them as uniform copies. Workers in each government bureaucracy wear nearly similar styles of clothing. In Office of Records, ordinary workers all wear sleeveless sweaters and the supervisors such as Lowry and Kurtzmann wear the same color and style of suit. When he arrives in Information Retrieval Lowry is cautioned by Jack Lint that "you'll never get anywhere in a suit like that," and gives him one identical to that worn by every other Information Retrieval officer.

In sum, recurring images in *Brazil* show how much the society is wrapped up in a mechtech aesthetic. Furthermore, that aesthetic serves the ends of fascism, repression, and violence. The population is cowed and terrified, or enlisted to conform and cooperate, not just through violence and regimentation but also through an aesthetic of violence and regimentation taken from mechtech. Now we turn to some ways in which electrotech aesthetics serve the same rhetorical functions.

FASCISM, SIMULATIONS, AND ELECTROTECH

There is less electrotech aesthetic in *Brazil* than there is mechtech or chaotech. But it is nevertheless present, and likewise supports a simulational fascist culture. Electrotech devices appear here and there: robots sniff and peer at people entering the Ministry of Information, plastic surgeons wear electric lights on their foreheads to examine patients, and Jack Lint relieves the stress of torturing people by rubbing his temples with buzzing electric devices. We noted earlier the ubiquity of television sets, which are of course electrotech. Harry Tuttle sneaks into Lowry's apartment for his first repair job on Lowry's heating system while calling him on the telephone. As they continue to talk, Tuttle walks up behind Lowry with a gun on him, still talking by phone although they are in the same room, preferring electrotech mediation over direct talk.

The central electrotech device in this society, as in ours, is the computer—despite the 1930s and 1940s clothing and decor. They seem to be everywhere and to operate as they do in any "real" late twentieth century society. The centrality of the computer is expressed in a slogan that appears on billboards: "Information: The Key to Prosperity." Of course "information" is *the* key term in computer parlance. We see this motto first on a huge sign in the Ministry of Information, and the link to simulational fascism is the Gestapo-like guard standing just beneath that sign, explaining the uses of a nine millimeter submachine gun to a nun. The link is made in a more ongoing way by the fact that the Ministry of Information is clearly the dominant power in this society.

Lowry is something of a computer expert, and has made himself invaluable to Mr. Kurtzmann in Records for his ability to use them. Also, some key plot twists such as his transfer to Information Retrieval have to do directly with computers: it is only as an Information Retrieval officer that Lowry can get the information he needs to find Jill Layton. In *Brazil* as in our society, many people are befuddled by computers: "I'm something of a whiz on that thing. . . . Computers are my forte," boasts Lowry's new office neighbor, Harvey Lime; but in reality he doesn't even know how to turn the machine on.

We learned that mystery and occlusion of the inner workings of electrotech machines is central to their aesthetic, as is the epiphany of mastering those workings. That same kind of mystery and occlusion is true of the film's society in general. It may be said that the bureaucracy of *Brazil* works like an electrotech machine, hiding itself from inquiry and empowering those who overcome those barriers.

Few people understand how the bureaucracy works. A check arrives in Records to reimburse Archibald Buttle for his expense in being arrested. But Buttle is dead from his torture. What to do with this simple piece of bureaucratic paper completely confounds Mr. Kurtzmann, who has to turn to Lowry (who is *also* his computer expert) for competent advice on how to make this social machine work. The complications of the society are likewise reflected in its electrotech machines. The telephones of *Brazil*, for instance, are overly complicated devices with little switchboards attached to them: just to answer the telephone one must pull and push plugs rapidly into place. In another example of occlusion, Lowry sees Layton pick up a package at the power plant and is consumed by curiosity as to what it is, fearing it is a bomb. Like an electrotech machine, its contents are hidden from his understanding. As with mechtech, the most electrotech machines, the most intense electrotech aesthetic, are concentrated in seats of the greatest fascist power. Arriving on the mysterious fiftieth floor, where Jack Lint's office and torture chamber are located, Sam finds a white-labcoated technician has removed a wall panel and is adjusting the electrotech switches, tape reels, and circuitry behind it for some nefarious reason. In the Office of the Deputy Minister of Information, Lowry finds row upon row of computer screens and printers spewing out arrest warrants.

Other evidence of electrotech aesthetics may be found. We noted in chapter three the androgyny of electrotech. One way to define androgyny is in terms of conventional expectations for gendered roles. In *Brazil*, the bureaucrats, who hold ultimate power, are nevertheless far from macho in a conventional sense. Early in the film, a bug drops into the typing mechanism of a computer, changing Tuttle to Buttle, because a bureaucrat in a white lab coat has climbed all over his office swatting at it; the man struts and preens triumphantly after finally hitting it, in a sort of parody of the little tailor who killed the flies. It is such a ridiculous false-macho gesture that it actually calls his strength, his manhood into question. Mr. Kurtzmann, Lowry's supervisor in Records, begs Lowry to sign some questionable papers in his stead, explaining that "my wrist has gone all limp"— a conventional reference to homosexuality, which is sometimes understood as entailing some degree of androgyny. In these instances, conventional standards for masculine behavior are violated by the male bureaucrats, suggesting an androgynous bent.

The fascist society of *Brazil* could not survive were it not for the computers and other electrotech devices that can be found everywhere. The aesthetic these devices foster supports the simulations that are central to that fascism. But the message of *Brazil* is also that fascism ultimately fails; the vehicle for that assertion is often a chaotech aesthetic.

FASCISM, SIMULATIONS, AND CHAOTECH

The audience for *Brazil* is, in general, the same sort of consumers that populate the film itself. We enjoy machine aesthetics as much as the next bureaucrat; so much so that we, like many other societies before us, might not mind if mechtech and electrotech are furthering the ends of a police state. So the film runs a rhetorical risk of depicting a society some might actually find attractive. To avoid that pitfall, the film presents *Brazil* as a rotting and ridiculous culture. It does that through chaotech, the aesthetic of the decayed machine, both mechtech and electrotech. This happens through depiction of "actual" chaotech, in machines that do not work and a city that is crumbling. Playing off of the metaphor of society as machine, chaotech aesthetic is also contained in the ways in which the society is pictured as dysfunctional, as a broken machine. The chaotech in *Brazil* is largely an ironic device: the audience is aware of decay and disaster, but the people in the film are less cognizant of it. As we shall see later in this chapter, nearly everyone in *Brazil* goes to great lengths to maintain the simulation even in the face of chaotech evidence to the contrary.

Images of old, decayed, malfunctioning, and dysfunctional machines abound in *Brazil*. Although television sets are strewn throughout the interiors of the film, they are the old fashioned 1950s kind inside large boxy consoles. One of the first images of the film is of a bank of these ancient, cranky sets in a store window. But then a terrorist bomb explodes, creating a chaotic/chaotech scene of destruction out of what was neat rows of high technology. As if to revel in this destruction, the camera pans in on one remaining set that, despite its smoking cabinet and sputtering, flickering screen still works more or less. It is on this decayed machine that we see an interview with the Deputy Minister of Information, promising viewers ironically that they are winning the war on terrorists. The machinery of that office clearly works no better than does the machinery of this bombed television.

Another remarkable example of chaotech aesthetic lies with the computers themselves. Although they are computers, with powers equal to any available to the audience, their aesthetic borders on chaotech. Screens are cheesey, grainy, large magnifying glasses, essentially; images are distorted and sometimes blurry. The "keyboards" for these machines are recognizable to the audience as taken from ancient manual typewriters; one must pound away hard to make them work. The look of these electrotech devices will be perceived by the film's audience, but likely not by the characters in the film, as evidence of cheap, flimsy, old fashioned construction on the verge of collapse and failure. That a mere fly dropping into a printer could cause the serious malfunction early in the film, in which Tuttle is changed to Buttle, is further evidence of the parlous state of this society's technology.

Lowry's apartment is a screen upon which progressively worse layers of chaotech are shown. At first we see he has what should be a marvelous, automated apartment in which everything is done for him by machines. But they don't work: a machine makes the morning coffee but then pours it onto the automated toast, which has likewise been a little burnt. That night Lowry awakes to a failed heating system: his apartment has drastically overheated. Renegade repairman Tuttle removes a panel in Lowry's wall and a chaotech jumble of pipes, ducts, and hoses comes tumbling out—as Tuttle is talking about the confusion and failure of Central Services' bureaucracy. The equation between failed government and failed machinery is clear. This scene foretells a greater chaos yet to come. Central Services enters the apartment while Lowry is out, and upon his return he finds his home filled with huge pipes, ducts, and hoses dangling from the ceiling and escaping from the walls. This confusion is echoed by the Central Services worker's taunting Lowry with the name "terrorist"—that powerful label of social dysfunction and decay. The last scene in his apartment a day later shows even worse damage, as Central Services has turned this model apartment into a deep freeze piled high with broken machinery and festooned with the ubiquitous ducts.

Other examples of failing machines are found throughout *Brazil*. Lowry has overslept one morning and is awakened by a telephone call from the frantic Mr. Kurtzmann. "The electrics here are up the spout," Lowry explains, and then after hearing Kurtzmann's reply continues, "Oh, yours too." When the police entered the Buttle apartment they came through a circular hole sawed in the ceiling (as they apparently do for every entry into a home). Hot on their heels is Central Services with a precut plug to put in the hole. But alas! it doesn't fit, for as one repair person complains to the other: "Typical, they've gone back to metric without telling us." Later, Lowry visits the Buttle apartment to give the refund check to Mrs. Buttle. He returns outside to find his car, which was a ridiculous tiny, flimsy contraption to begin with, has been burned, stripped, and put up on blocks. A similar outcome awaits two highly mechtech police cars that are chasing Lowry and Layton's truck. These cars, huge massive hunks of steel parts, turrets, exposed bolts and rivets, and other mechtech accoutrements, run into the load that Lowry and Layton drop off of their truck, exploding spectacularly in an orgy of destruction.

Even (or perhaps, especially) in the all powerful Ministry of Information, machines do not work as they should. Lowry has elevator trouble: going to his new job for the first time he finds that the elevator does not stop even with the floor and he must climb out of it. Later, he takes the elevator to the basement just in time for repair people to hang a "Lift Out Of Order" sign on it, although it has obviously functioned well enough to take him there. Toward the end of the movie, Lowry slumps in despair in his office as paper piles higher and higher. Exasperated and unable to keep up with the flow of tasks that come to him in cylinders through pneumatic tubes in his office, Lowry sabotages the system by connecting the in and out tubes. This causes an explosion of the entire system, and he and other workers step out into the hallway to see a snowshower of paper coming down throughout the building from the burst ductwork in the ceiling.

If the society in *Brazil* is to run like a machine, then evidence of collapse and decay in its organization, especially in the bureaucracy, is a sort of social chaotech to parallel the chaotech afflicting the "real" machines. This equation is made most explicit when Central Services workers first arrive at Lowry's apartment to repair his dysfunctional heating system. Lowry cannot allow them in, for Harry Tuttle is lying in wait for them with a gun. He puts them off by asking if they have the proper form: a twenty-seven B stroke six. They do not, and are stymied; this failure of the bureaucratic machine, which prevents them from entering the apartment to fix the failed machinery of the heating apparatus, causes one of the workers to begin shaking, shuddering, and clattering exactly like a broken machine himself. The broken bureaucracy is reflected in broken machines and in images of machines, and the broken worker-as-machine has to be led off by his partner.

Examples of what one might call social chaotech, based on the idea of this society as a failing machine, occur often in the film. Machine aesthetics feature efficiency and production, and the bureaucracy of *Brazil* is devoid of either. It is clear to the audience that the society is drowning in paperwork. Only a few characters in the film itself are aware of how the bureaucracy is failing. When asked by Lowry why he no longer works for Central Services, Tuttle replies, "I couldn't stand the paperwork," and then launches into a diatribe against it. Yet Tuttle himself has now become a broken part of that apparatus, working against its normal functioning. Lowry is also a dysfunctional part in this machine because he has consistently refused promotion from Records. His friends do not understand this lack of ambition in a society that values nothing but climbing the bureaucratic ladder, but he doesn't care: "I know. Wonderful, marvelous, perfect," he replies to charges that Records is a dead end job. Layton is a broken chaotech cog in the machine because she makes a pest of herself by questioning the Buttle arrest. Pursued by the police, a terrified Lowry jumps into Layton's truck and screams at her, "Drive, trust me, you are in terrible danger, you are an embarrassment." That final odd word is telling, for it promotes the social dysfunction of embarrassment from personal faux pas to grave physical danger. Layton's refusal to go along with the rest of the machine has created a social dysfunction that threatens to derail the mechanism.

The angst and terror generated by the refund check for Buttle's torture also illustrates social chaotech. The practice in this society is to dun the bank accounts of torture victims to make them pay for their own inquisitions. When it is discovered that Buttle was falsely arrested, a check is issued to reimburse him for the inconvenience, although he is dead by that time. This has evidently never happened before, and it throws Mr. Kurtzmann, of Records, to whom the check and accompanying paperwork come, into an absolute panic: "Well, that's it. I'd just as well go hang myself now." The machinery of bureaucracy has had a wrench thrown into it. Lowry does what he can to suggest ways to resolve the matter, but it is clear that all his solutions are jerry-rigged repairs on a machine that is not designed to admit its own mistakes.

The rhetorical appeals of *Brazil* to its audience are thus furthered by chaotech images of real machines and of a kind of social chaotech befalling this society that imagines itself as a machine. Chaotech gives the lie, for the sake of the audience in the theater, to the fascist culture of *Brazil* which is wrapped in its simulations based on mechtech and electrotech aesthetics. But the warnings offered to the audience by chaotech do not seem to affect the people of *Brazil*. The hold of the simulation is so strong that they cannot break free. That, too, is part of the message of the film: Simulations are highly seductive and difficult to escape once one is inside them. We examine the durability of these simulations in the next section.

TRAPPED INSIDE SIMULATIONS

Every now and then raw, brutal experience moves in upon the simulations enjoyed by the characters in *Brazil*, threatening to rip apart their aestheticized lives and expose the real terror and tawdriness of that society. Glimpses of "the real," if one may use such an old fashioned concept, break through from time to time. The Buttle home, despite the ubiquitous ductwork running through it, is really quite old fashioned and nonmechanical. This domestic scene is destroyed by the truly terrifying attack on it by black uniformed, helmeted assault troops. Later, Lowry visits the ruined apartment to return the check to Mrs. Buttle. His bureaucratic prattle is stopped cold by her anguished, grief stricken cry, "What have you done with his body?" Fleeing from the police in Layton's lorry, Lowry manages to dump the truck's load so that the pursuing cars crash into it. Lowry shrieks with joy, "We did it!" until he sees the reality of what they have done: burning police officers stagger from the wreckage in their death throes. "Oh no!" he whispers, shocked at the reality of what he has done. Finally, the terrorist bombings during the film create real pain and suffering for those who are injured, if for nobody else. Bloodied and broken bodies cry out in pain amidst the wreckage.

Despite these momentary rendings of the veil, the hold of simulation is so powerful on the characters in *Brazil* that it becomes self sealing, self healing, self correcting. Attempts to escape simulation inevitably fall back into it. Alternatives to an aestheticized life become aestheticized in turn. In this way, the film warns its audience about the seductive dangers of simulation. To understand this, let us consider alternatives to the machine aesthetic simulation that are constructed based on privilege and power, on marginalization, and on love.

One alternative to the fascist simulations of *Brazil* is to have power and influence. Lowry's mother is in that position, and is able to escape much, although not all, of the machine aesthetic simulations and thus, the fascist oppression, of the society. But she and her friends accomplish this at the cost of alternative simulations. Mrs. Lowry and her cronies are obsessed with plastic surgery; they go under the knife constantly, with astonishingly rejuvenating results—except for one unfortunate woman whose incompetent doctor is using a controversial acid treatment, which causes her to be wrapped in more and more bloody bandages as the film progresses. She explains, "I had a little complication," and then, "My complication had a little complication," to Lowry as he watches her slow physical decay in horror. The wealthy women of *Brazil* use their own bodies as a canvas of aestheticization, of simulation.

The architecture and decor of the privileged also shuns a machine aesthetic, only to embrace alternative simulations. The plastic surgeons' offices are sumptuously appointed in nineteenth century grandeur, not at all in a machine aesthetic style. Mrs. Lowry's home looks like an Edwardian palace. Her servants are dressed as eighteenth century footmen although they carry pistols and metal detecting wands. The restaurants these wealthy people go to are Art Nouveau spectacles.

In short, the way that the rich and powerful of this society "escape" from the simulations of their culture is to go to alternative simulations. Not even the intrusion of terrorist explosions can break in on this make-believe: The trio in the bombed restaurant picks itself up and, smudged and disheveled, returns to playing chamber music for the benefit of the surviving patrons. Screens are hurriedly put up to shield the diners from the scene of carnage not thirty feet from them, and they continue their merrymaking.

Another alternative to fascist simulations is made available to the marginalized in *Brazil*, specifically the poor and terrorists. We do not see many poor people, but they appear from time to time, principally in the Buttle's neighborhood. These people cannot afford the machines and grand, geometric architecture of most other citizens. They live in chaotech most of the time, if not in downright squalor. In case their estrangement from *Brazil*'s simulations is not clear enough to the film's audience, the word "reality" is written in large letters on an alley wall as Lowry drives up to the Buttle's building. Yet even these poor have their simulations as alternatives to their wretched estrangement from the film's shiny machine aesthetics. Just before his arrest, Buttle and family were lost in the enjoyment of television, like everyone else. As Lowry goes off to find Mrs. Buttle, we see that the children of the poor are playing out fantasies of arrest and torture themselves. Clearly the poor yearn for the dominant simulations of their culture, even to participate in violence and brutality themselves, if only they could afford it.

The terrorists in the film are also locked in their own simulations even as they violently oppose the fascist simulations of the society. *What* gets bombed is significant here. The first terrorist target we see is a television store; the second is the fancy restaurant in which the Lowrys are dining; the third is a nightclub called "The Blue Lagoon"; and the fourth target is a department store. The real centers of power such as the energy plant outside the city go unmolested. The Ministry of Information is never touched by violence (except in Lowry's final fantasy). What kind of terrorists are these who go after television stores and bars but not police stations? Lowry asks Tuttle much the same question, demanding to know why he works as a renegade repairman. No political answer is returned, no refusal of the fascist simulation. Instead: "I came into this game for the action, the excitement, go anywhere, travel light." By way of illustrating this jocular sense of mission, Tuttle's final act of revenge against Central Services is to switch air and sewer hoses as two engineers work inside space suits in Lowry's apartment. Their suits fill with sewage to the delight of Lowry and Tuttle, viewing the scene from outside.

Those who are marginalized in this society, who do not participate in and do not benefit from the simulational society based on machine aesthetics, nevertheless have simulational fantasies of their own. Like the audience's "own" terrorists who mug for the CNN cameras, those who refuse fascism in *Brazil* do so at least in part because it enables them to pursue exciting and amusing alternative simulations.

One final alternative to the fascist simulations of *Brazil* is love, and the chief proponent of that alternative is Sam Lowry. His love for Layton is irrational, it does not follow proper procedure, it violates all the rules. Love itself goes against the logic of *Brazil*, for it refuses bureaucracy and rigid technique. From the beginning, Lowry's motivation of love gets him into deep trouble, and it is eventually his undoing.

But the way Lowry pursues this love, the way he makes it manifest, the way it works for him in his life, is every bit as simulational as is the culture he is fleeing. From time to time, in his dreams, daydreams, or when nodding in exhaustion at his desk, Lowry enters into his own fantasy world in which he is a creature with mechanical wings flying to the rescue of the fair Layton who is trapped in an iron cage, held there by huge monsters. This simulation is clearly an alternative to Lowry's drab life in the simulational culture of *Brazil*; but he has only traded one aestheticized reality for another.

Furthermore, Lowry's choice of realities seems to be to bounce back and forth between the simulations of *Brazil* and his dream fantasy. He is able to slip into the fantasy while at work and even in public places with ease. The fantasy works as a simulation for him, though. At one point he is asked if he has no ambition, no dreams, and he replies, "No, not even dreams." If the fantasy is not that, then it must be a simulation, for it is surely not reality.

The borders between one simulation and another become increasingly porous for Lowry as the film progresses. The uniformed soldier who is about to arrest Layton becomes, in Lowry's eyes, the monster of his fantasy, which he attacks and subdues. Lowry leaves Layton in his mother's apartment to break into the Ministry of Information and alter her computer records. He returns to inform her, "You don't exist anymore. I've killed you. Jill Layton is dead." There is a simulation *somewhere* here, since Layton stands living before him, but the audience and perhaps even Lowry cannot tell which version of their experience it is. There are other examples of the merging of simulations in Lowry's life. In his fantasy, Layton has long, flowing hair; in the simulational culture of *Brazil*, it is extremely short. Yet when Lowry finally makes love to her in his mother's apartment, Layton appears (to both Lowry and to the audience) to have long hair. When they wake up the next morning to the sound of the police bursting in, it is short again.

Lowry's final simulation is to imagine his rescue by terrorists led by Harry Tuttle as he sits in the torture chamber. The audience, which has been increasingly unable to tell what are Lowry's dream fantasies and what are his "reality" is taken in by this one: we, too, thrill to the sight of armed terrorists shooting the torturer in the forehead and sliding down ropes to Lowry's rescue, whisking him away, running to escape the police. But the scenario begins to grow a bit strange, as Lowry runs into the by now dead Layton disguised as his mother attending a surreal funeral for his mother's friend. The final scene shows Lowry and Layton escaping into a pastoral, sylvan scene that no longer exists in the industrialized world of *Brazil*—and then the scene changes and we see Lowry's stark mad, staring eyes as he sits in the torture chair to the lament of his inquisitors, "We lost him!"

In short, Lowry's alternatives to the simulations of *Brazil* are simulational themselves, and become even more so as the film progresses. These are powerful simulations, and that is part of the message for the audience as well, for we are sucked up into them ourselves from time to time. Our ability to sit in judgment of the culture of *Brazil* is shown to be on weak ground; we love a good rescue scene, a little movie romance, as much as anyone. We are ready and willing to enter into Lowry's simulation with him.

The rhetorical message of *Brazil* is therefore not merely one of warning its audience against simulations that prop up fascism. Those simulations, as we have seen, are based largely on machine aesthetics. But the audience viewing the film is having an aesthetic experience of its own, and likely enjoying it immensely. Although watching the film in and of itself is not likely to further fascist tendencies, the inability of the characters to escape simulation speaks powerfully to an audience that is wrapped up in the simulation that is the film *Brazil*. We are told in essence that this could happen to us, it could be us. An audience is invited then to interrogate its own love of simulation, and to consider the political consequences thereof.

CONCLUSION

Machine aesthetics is an experience and a subject for inquiry that is vast and ongoing. This book should be considered, not as the final word on either types of machine aesthetic or their dimensions, but as encouragement for further investigation into the matter. For instance, it may be that there is a distinct aesthetic of the *recycled* machine; not a decayed one, but one that has been reclaimed, refurbished, and sent upon new missions.

More might be done with extending a machine aesthetic, or perhaps more accurately a *sensibility* based on a machine aesthetic, into nonmechanical areas of life. How do metaphors based on machines work? If a government, a culture, an office, or a household imagines itself as a machine, or conducts itself along lines of certain dimensions of machine aesthetic, what effect does that have on such conduct?

Clearly, the rhetorical effects of machine aesthetics in particular needs development. What I have attempted to do here is to lay out a sort of logic or set of parameters for how machines are experienced aesthetically and what those aesthetics might mean. The rhetorical applications of such aesthetics are a wide and fruitful field for study.

We live our lives with, around, and sometimes against machines. They come to represent our hopes as well as our fears. We use them, but we also experience them aesthetically. I have tried to explain those aesthetics systematically, and to show how the aesthetic reactions might be used rhetorically.

Works Cited

Aldersey-Williams, Hugh, ed. *New American Design: Products and Graphics for a Post-Industrial Age*. New York: Rizzoli, 1988.
Aldiss, Brian W. "Robots: Low-Voltage Ontological Currents." *The Mechanical God: Machines in Science Fiction*. Ed. Thomas P. Dunn and Richard D. Erlich. Westport, CT: Greenwood P, 1982. 3–9.
Altheide, David L. *Media Power*. Beverly Hills, CA: Sage, 1985.
Aristotle. *Rhetoric and Poetics of Aristotle*. Trans. W. Rhys Roberts and Ingram Bywater. New York: Modern Library, 1954.
Arnheim, Rudolf. *Entropy and Art: An Essay on Disorder and Order*. Berkeley: U of California P, 1971.
Asimov, Morris. *Introduction to Design*. Englewood Cliffs, NJ: Prentice-Hall, 1962.
Barlow, John Perry. "Crime and Puzzlement." *CyberReader*. Ed. Victor Vitanza. Boston: Allyn & Bacon, 1996. 92–115.
Barre, François. "Design in Question." *Industrial Design: Reflection of a Century*. Ed. Jocelyn de Noblet. Paris: Flammarion, 1993. 10–11.
Baudrillard, Jean. *Simulations*. Trans. Paul Foss, Paul Patton, and Philip Beitchman. New York: Semiotext(e), Inc., 1983.
———. *The System of Objects*. Trans. James Benedict. London: Verso, 1996.
Benford, Gregory. "Science Fiction, Rhetoric, and Realities: Words to the Critic." *Fiction 2000: Cyberpunk and the Future of Narrative*. Ed.

George Slusser and Tom Shippey. Athens: U of Georgia P, 1992. 223–29.

Birkerts, Sven. "Into the Electronic Millennium." *CyberReader*. Ed. Victor Vitanza. Boston: Allyn & Bacon, 1996. 203–14.

Bonner, Frances. "Separate Development: Cyberpunk in Film and TV." *Fiction 2000: Cyberpunk and the Future of Narrative*. Ed. George Slusser and Tom Shippey. Athens: U of Georgia P, 1992. 191–207.

Bright, Deborah. "The Machine in the Garden Revisited: American Environmentalism and Photographic Aesthetics." *Art Journal* 51.2 (Summer 1992): 60–71.

Brolin, Brent C. *The Failure of Modern Architecture*. New York: Van Nostrand, 1976.

Brummett, Barry. *Contemporary Apocalyptic Rhetoric*. Tuscaloosa: U of Alabama P, 1991.

Bryant, Donald C., ed. *Papers in Rhetoric and Poetic*. Iowa City: U of Iowa P, 1965.

Burke, K. *Attitudes Toward History*. 2 vols. New York: New Republic, 1937.

———. *A Grammar of Motives*. New York: Prentice-Hall, 1945.

———. *The Philosophy of Literary Form: Studies in Symbolic Action*. Baton Rouge, LA: Louisiana State UP, 1941.

———. *A Rhetoric of Motives*. New York: Prentice-Hall, 1950.

Burr, Arthur H., and John B. Cheatham. *Mechanical Analysis and Design*. 2nd ed. Englewood Cliffs, NJ: Prentice-Hall, 1995.

Caplan, Ralph. "Preface." *New American Design: Products and Graphics for a Post-Industrial Age*. Ed. Hugh Aldersey-Williams. New York: Rizzoli, 1988. 8–11.

Cheney, Sheldon, and Martha Cheney. *Art and the Machine*. New York: Whittlesey House, 1936.

Christie, John. "Of AIs and Others: William Gibson's Transit." *Fiction 2000: Cyberpunk and the Future of Narrative*. Ed. George Slusser and Tom Shippey. Athens: U of Georgia P, 1992. 171–82.

Clark, Nigel. "Rear-View Mirrorshades: The Recursive Generation of the Cyberbody." *Cyberspace/Cyberbodies/Cyberpunk*. Ed. Mike Featherstone and Roger Burrows. London: Sage, 1995. 113–33.

Coleman, Francis J., ed. *Contemporary Studies in Aesthetics*. New York: McGraw-Hill, 1968.

Crowther, Paul. *Critical Aesthetics and Postmodernism*. Oxford: Clarendon, 1993.

Csicsery-Ronay, Istvan, Jr. "Futuristic Flu, or, The Revenge of the Future." *Fiction 2000: Cyberpunk and the Future of Narrative*. Ed. George Slusser and Tom Shippey. Athens: U of Georgia P, 1992. 26–45.

Davidson, Cynthia C., ed. *Anyway*. New York: Rizzoli, 1994.

Del Caro, Adrian. *Dionysian Aesthetics: The Role of Destruction in Creation as Reflected in the Life and Works of Friedrich Nietzsche*. Frankfurt am Main: Peter D. Lang, 1981.

DeLeuze, Gilles and Felix Guattari. *Anti-Oedipus: Capitalism and Schizophrenia*. Minneapolis: U of Minnesota P, 1983.
Dewey, John. *Art as Experience*. New York: Minton, Balch, 1934.
Diller, Elizabeth. "Bad Press: Housework Series." *Anyway*. Ed. Cynthia C. Davidson. New York: Rizzoli, 1994. 153–61.
Downey, Gary Lee. "The World of Industry-University-Government: Reimagining R&D as America." *Technoscientific Imaginaries: Conversations, Profiles, and Memoirs*. Ed. George Marcus. Chicago: U of Chicago P, 1995. 197–226.
Dunn, Thomas P. and Richard D. Erlich, eds. *The Mechanical God: Machines in Science Fiction*. Westport, CT: Greenwood P, 1982.
Eidelberg, Martin, ed. *Design 1935-1965: What Modern Was*. New York: Harry N. Abrams, Inc., 1991.
Ellul, Jacques. *The Technological Society*. Trans. John Wilkinson. New York: Alfred A. Knopf, 1964.
Endt, Evert, and Sabine Grandadam. "Design for Everyday Objects." *Industrial Design: Reflection of a Century*. Ed. Jocelyn de Noblet. Paris: Flammarion, 1993. 30–36.
Ewen, Stuart. *All Consuming Images: The Politics of Style in Contemporary Culture*. New York: Basic, 1988.
Ewen, Stuart, and Elizabeth Ewen. *Channels of Desire: Mass Images and the Shaping of American Consciousness*. Minneapolis: U of Minnesota P, 1992.
Featherstone, Mike, and Roger Burrows, eds. *Cyberspace/Cyberbodies/Cyberpunk*. London: Sage, 1995.
Ferebee, Ann. *A History of Design From the Victorian Era to the Present*. New York: Van Nostrand Reinhold, 1970.
Florman, Samuel C. *The Existential Pleasures of Engineering*. 2nd. ed. New York: St. Martin's, 1994.
Fogg, Walter L. "Technology and Dystopia." *Utopia/Dystopia?* Ed. Peyton E. Richter. Cambridge, MA: Schenkman, 1975. 57–73.
Frank, Felicia Miller. *The Mechanical Song: Women, Voice, and the Artificial in Nineteenth-Century French Narrative*. Stanford, CA: Stanford UP, 1995.
Gelerntner, David. *Machine Beauty: Elegance and the Heart of Technology*. New York: Basic Books, 1998.
Giedion, Siegfried. *Mechanization Takes Command*. 1948. New York: W. W. Norton, 1969.
Gilliam, Terry, dir. *Brazil*. With Jonathan Pryce and Robert De Niro. Embassy International Pictures, 1985.
Greenhalgh, Paul, ed. *Modernism in Design*. London: Reaktion Books, Ltd., 1990.
Gropius, Walter. *The New Architecture and the Bauhaus*. Trans. P. Morton Shand. Cambridge, MA: MIT UP, 1965.

Guillerme, Jacques. "Design in the First Machine Age." *Industrial Design: Reflection of a Century.* Ed. Jocelyn de Noblet. Paris: Flammarion, 1993. 53–61.

Hafner, Katie, and John Markoff. *Cyberpunk: Outlaws and Hackers on the Computer Frontier.* New York: Touchstone, 1995.

Hahn, Roger. "The Meaning of the Mechanistic Age." *The Boundaries of Humanity: Humans, Animals, Machines.* Ed. James J. Sheehan and Morton Sosna. Berkeley: U of California P, 1991. 142–57.

Hamm, Manfred. *Dead Tech: A Guide to the Archaeology of Tomorrow.* San Francisco: Sierra Club Books, 1981.

Hampshire, Stuart. "Biology, Machines, and Humanity." *The Boundaries of Humanity: Humans, Animals, Machines.* Ed. James J. Sheehan and Morton Sosna. Berkeley: U of California P, 1991. 253–56.

Hancock, Marion. "Industrial Design Today." *Industrial Design: Reflection of a Century.* Ed. Jocelyn de Noblet. Paris: Flammarion, 1993. 268–77.

Happold, Edward. "Can You Hear Me at the Back?" *The Structural Engineer.* 64A (1986): 367–78.

Haraway, Donna J. *Simians, Cyborgs, and Women: The Reinvention of Nature.* New York: Routledge, 1991.

Harper's Forum. "Is Computer Hacking a Crime?" *CyberReader.* Ed. Victor Vitanza. Boston: Allyn & Bacon, 1996. 72–91.

Heim, Michael. "The Design of Virtual Reality." *Cyberspace/Cyberbodies/Cyberpunk.* Eds. Mike Featherstone and Roger Burrows. London: Sage, 1995. 65–77.

———. "The Essence of VR." *CyberReader.* Ed. Victor Vitanza. Boston: Allyn & Bacon, 1996. 16–30.

Hine, Thomas. *Populuxe.* New York: Alfred A. Knopf, 1986.

Holgate, Alan. *Aesthetics of Built Form.* Oxford: Oxford UP, 1992.

Howell, Wilbur Samuel. *Poetics, Rhetoric and Logic: Studies in the Basic Disciplines of Criticism.* Ithaca: Cornell UP, 1975.

Jackson, Lesley. *The New Look: Design in the Fifties.* New York: Thames and Hudson, 1991.

Jameson, Fredric. "The Uses of Apocalypse." *Anyway.* Ed. Cynthia C. Davidson. New York: Rizzoli, 1994. 33–41.

Kantrowitz, Barbara. "Men, Women, Computers." *CyberReader.* Ed. Victor Vitanza. Boston: Allyn & Bacon, 1996. 134–40.

Kaufer, David S., and Brian S. Butler. *Rhetoric and the Arts of Design.* Mahwah, NJ: Lawrence Erlbaum Associates, 1996.

Kirby, Michael. "The Aesthetics of the Avant-Garde." *Esthetics Contemporary.* Ed. Richard Kostelanitz. Buffalo: Prometheus Books, 1978. 36–70.

Kostelanitz, Richard, ed. *Esthetics Contemporary.* Buffalo: Prometheus Books, 1978.

Kouwenhoven, John A. *Half a Truth Is Better Than None: Some Unsystematic Conjectures About Art, Disorder, and American Experience.* U of Chicago P, 1982.
Kwinter, Sanford. "The Complex and the Singular." *Anyway.* Ed. Cynthia C. Davidson. New York: Rizzoli, 1994. 186–97.
Landow, George. *Hypertext: The Convergence of Contemporary Critical Theory and Technology.* Baltimore: Johns Hopkins UP, 1992.
Lanham, Richard A. *The Electronic Word: Democracy, Technology, and the Arts.* Chicago: U of Chicago P, 19.
Larijani, L. Casey. *The Virtual Reality Primer.* New York: McGraw-Hill, 1994.
Lauretis, Teresa de, Andreas Huyssen, and Kathleen Woodward, eds. *The Technological Imagination: Theories and Fictions.* Madison, WI: Coda, 1980.
Le Corbusier. *Toward a New Architecture.* 1927. Trans. Frederick Etchells. London: The Architectural P, 1963.
Léger, Fernand. "The Esthetics of the Machine." *The Little Review.* 9.3 (Spring 1923): 45–49.
Lupton, Deborah. "The Embodied Computer/User." *Cyberspace/Cyberbodies/Cyberpunk.* Ed. Mike Featherstone and Roger Burrows. London: Sage, 1995. 97–112.
Marcus, George H. *Functionalist Design: An Ongoing History.* New York: Prestel, 1995.
———, ed. *Technoscientific Imaginaries: Conversations, Profiles, and Memoirs.* Chicago: U of Chicago P, 1995.
Marcuse, Herbert. *One-Dimensional Man: Studies in the Ideology of Advanced Industrial Society.* Boston: Beacon, 1964.
Mayall, W. H. *Machines and Perception in Industrial Design.* London: Studio Vista/Reinhold, 1968.
McClintock, Alexander. *The Convergence of Machine and Human Nature: A Critique of the Computer Metaphor of Mind and Artificial Intelligence.* Aldershot, UK: Avebury, 1995.
McCoy, Michael and Katherine McCoy. "Introduction." *New American Design: Products and Graphics for a Post-Industrial Age.* Ed. Hugh Aldersey-Williams. New York: Rizzoli, 1988.
McGuirk, Carol. "The 'New' Romancers: Science Fiction Innovators From Gernsbach to Gibson." *Fiction 2000: Cyberpunk and the Future of Narrative.* Ed. George Slusser and Tom Shippey. Athens: U of Georgia P, 1992. 109–29.
McKeon, Richard. *Rhetoric: Essays in Invention and Discovery.* Ed. Mark Backman. Woodbridge, CT: Ox Bow P, 1987.
McLuhan, Marshall. *Understanding Media: The Extensions of Man.* 2nd ed. New York: McGraw-Hill, 1964.

McLuhan, Marshall, and Quentin Fiore. *The Medium Is the Message: An Inventory of Effects.* New York: Bantam, 1967.
Meikle, Jeffrey L. "Streamlining 1930–1955." *Industrial Design: Reflection of a Century.* Ed. Jocelyn de Noblet. Paris: Flammarion, 1993. 182–92.
Mills, Sara, ed. *Language and Gender: Interdisciplinary Perspectives.* New York: Longman, 1995.
Moholy-Nagy, Sibyl. *Moholy-Nagy: Experiment in Totality.* Cambridge, MA: MIT UP, 1969.
Mumford, Lewis. *Art and Technics.* New York: Columbia UP, 1952.
———. *Technics and Civilization.* New York: Harcourt, Brace, 1934.
Newell, Allen. "Metaphors for Mind, Theories of Mind: Should the Humanities Mind?" *The Boundaries of Humanity: Humans, Animals, Machines.* Ed. James J. Sheehan and Morton Sosna. Berkeley: U of California P, 1991. 158–97.
Noblet, Jocelyn de, ed. *Industrial Design: Reflection of a Century.* Paris: Flammarion, 1993.
Novotny, Patrick. "No Future! Cyberpunk, Industrial Music, and the Aesthetics of Postmodern Disintegration." *Political Science Fiction.* Ed. Donald M. Hassler and Clyde Wilcox. Columbia, SC: U of S Carolina P, 1997. 99–123.
Nye, David E. *Electrifying America: Social Meanings of a New Technology, 1880–1940.* Cambridge, MA: MIT UP, 1990.
Olson, Kathryn M., and G. Thomas Goodnight. "Epochal Rhetoric in 19th-Century America: On the Discursive Instantiation of the Technical Sphere." *Spheres of Argument.* Ed. Bruce Gronbeck. Annandale, VA: SCA, 1989. 57–65.
Peckham, Morse. "Art and Disorder." *Esthetics Contemporary.* Ed. Richard Kostelanitz. Buffalo: Prometheus Books, 1978. 95–115.
———. *Man's Rage For Chaos: Biology, Behavior, and the Arts.* Philadelphia: Chilton, 1965.
Peirce, Charles S. *Charles S. Peirce: The Essential Writings.* Ed. Edward C. Moore. New York: Harper & Row, 1972.
Perkowitz, Sidney. "Hubs, Struts, and Aesthetics." *Technology Review* 99 (November/December 1996): 56–63.
Plant, Sadie. "The Future Looms: Weaving Women and Cybernetics." *Cyberspace/Cyberbodies/Cyberpunk.* Ed. Mike Featherstone and Roger Burrows. London: Sage, 1995. 45–64.
Plato. *Gorgias.* Trans. Walter Hamilton. Baltimore: Penguin, 1960.
Polanyi, Michael. *The Tacit Dimension.* Garden City, NY: Doubleday, 1966.
Poster, Mark. "Postmodern Virtualities." *Cyberspace/Cyberbodies/Cyberpunk.* Ed. Mike Featherstone and Roger Burrows. London: Sage, 1995. 79–95.

Pursell, Carroll. "The American Ideal of a Democratic Technology." *The Technological Imagination: Theories and Fictions*. Ed. Teresa de Lauretis, Andreas Huyssen, and Kathleen Woodward. Madison, WI: Coda, 1980. 11–25.
Rajchman, John. "Manyways." *Anyway*. Ed. Cynthia C. Davidson. New York: Rizzoli, 1994. 163–67.
Ramoneda, Josep. "Anyway: Geopolitics and Architecture." *Anyway*. Ed. Cynthia C. Davidson. New York: Rizzoli, 1994. 53–59.
Read, Herbert. *Art and Industry*. 1934. 5th ed. London: Faber & Faber, 1966.
Richter, Peyton E., ed. *Utopia/Dystopia?* Cambridge, MA: Schenkman, 1975.
Robins, Kevin. "Cyberspace and the World We Live In." *Cyberspace/Cyberbodies/Cyberpunk*. Ed. Mike Featherstone and Roger Burrows. London: Sage, 1995. 135–55.
Sanders, Joe. "Tools/Mirrors: The Humanization of Machines." *The Mechanical God: Machines in Science Fiction*. Ed. Thomas P. Dunn and Richard D. Erlich. Westport, CT: Greenwood P, 1982. 167–76.
Saper, Craig J. *Artificial Mythologies: A Guide to Cultural Invention*. Minneapolis: U of Minnesota P, 1997.
Sayre, Henry M. "American Vernacular: Objectivism, Precisionism, and the Aesthetic of the Machine." *Twentieth Century Literature* 35 (1989): 310–42.
Saussure, Ferdinand de. *Course in General Linguistics*. Ed. Charles Bally and Albert Sechehaye. Trans. Wade Baskin. London: Fontana, 1974.
Selzer, Michael. *Terrorist Chic: An Exploration of Violence in the Seventies*. New York: Hawthorn Books, Inc., 1979.
Setzer, Mark. *Bodies and Machines*. New York: Routledge, 1992.
Sheehan, James J., and Morton Sosna, eds. *The Boundaries of Humanity: Humans, Animals, Machines*. Berkeley: U of California P, 1991.
Sheppard, Anne. *Aesthetics: An Introduction to the Philosophy of Art*. New York: Oxford UP, 1987.
Shimer, Lewis. "Inside the Movement: Past, Present, and Future." *Fiction 2000: Cyberpunk and the Future of Narrative*. Ed. George Slusser and Tom Shippey. Athens: U of Georgia P, 1992. 17–25.
Slusser, George, and Tom Shippey, eds. *Fiction 2000: Cyberpunk and the Future of Narrative*. Athens: U of Georgia P, 1992.
Solà-Morales Rubió, Ignasi de, and Josep Ramoneda. "Anyway." *Anyway*. Ed. Cynthia C. Davidson. New York: Rizzoli International/Anyone Corporation, 1994. 20–23.
Tabbi, Joseph. *Postmodern Sublime: Technology and American Writing From Mailer to Cyberpunk*. Ithaca: Cornell UP, 1995.
Taylor, Mark C. "Net Working." *Anyway*. Ed. Cynthia C. Davidson. New York: Rizzoli, 1994. 99–105.

Tomas, David. "Feedback and Cybernetics: Reimaging the Body in the Age of the Cyborg." *Cyberspace/Cyberbodies/Cyberpunk*. Ed. Mike Featherstone and Roger Burrows. London: Sage, 1995. 21–43.

Turkle, Sherry. "Romantic Reactions: Paradoxical Responses to the Computer Presence." *The Boundaries of Humanity: Humans, Animals, Machines*. Ed. James J. Sheehan and Morton Sosna. Berkeley: U of California P, 1991. 224–252.

———. *The Second Self: Computers and the Human Spirit*. New York: Simon & Schuster, 1984.

Ullman, Ellen. *Close to the Machine: Technophilia and Its Discontents*. San Francisco: City Lights Books, 1997.

Urmson, J. O. "What Makes a Situation Aesthetic." *Contemporary Studies in Aesthetics*. Ed. Francis J. Coleman. New York: McGraw-Hill, 1968. 356–69.

Van Doren, Harold. *Industrial Design*. New York: McGraw-Hill, 1940.

Vitale, Elodie. "The Bauhaus and the Theory of Form." *Industrial Design: Reflection of a Century*. Ed. Jocelyn de Noblet. Paris: Flammarion, 1993. 154–164.

Vitanza, Victor, ed. *CyberReader*. Boston: Allyn & Bacon, 1996.

Washabaugh, William. *Flamenco: Passion, Politics and Popular Culture*. Oxford, UK: Berg, 1996.

Wiener, Norbert. *Cybernetics: Or Control and Communication in the Animal and the Machine*. 2nd ed. Cambridge, MA: MIT UP, 1961.

Whitson, Steve and John Poulakos. "Nietzsche and the Aesthetics of Rhetoric." *Quarterly Journal of Speech* 79 (1993): 131–45.

Wolmark, Jenny. "Cyborgs and Cyberpunk: Rewriting the Feminine in Popular Fiction." *Language and Gender: Interdisciplinary Perspectives*. Ed. Sara Mills. New York: Longman, 1995. 107–20.

Woolley, Benjamin. "Cyberspace." *CyberReader*. Ed. Victor Vitanza. Boston: Allyn & Bacon, 1996. 5–15.

Yeh, Susan Fillin. *Charles Sheeler and the Machine Age*. Ph.D. dissertation, City University of New York, 1981.

Zucker, Paul. *Fascination of Decay: Ruins: Relic-Symbol-Ornament*. Ridgewood, NJ: The Gregg P, 1968.

Index

Aesthetics, 3–10, 119–21; categories and schemes of, 10; defined, 9; faculties, 5–7, 14; properties, 7–10; rhetoric and, 23–24; systems, 5
African American, 45
Airplanes, 14, 38, 73–74, 83
America Online, 80
Anthropomorphism, 68, 72, 84
Apocalypse and apocalyptic, 90, 97–98, 106
Architecture, 32–33, 36–38, 42–43, 46, 53, 98–99, 123, 134
Art, 2, 5–6, 26, 91–92, 98
Automation, 41–43, 53, 81, 83, 109, 123. *See also* Robot
Automobiles, 34–35, 39, 47, 51, 59, 63, 89–90, 94, 96, 99, 102

Bauhaus, 31, 54, 99
Brazil, 3–4, 113–37

Burke, Kenneth, 8–10, 55, 86

Capitalism, 107–8, 110, 117, 119
Chaos, 90–92, 130
Chaotech, 4, 18–19, 28, 31, 83, 88–114, 118, 129–33
Class, 20, 48
Cleanliness, 37–39, 53, 71–72
Clocks, 2, 34, 42–43, 46, 48, 52, 97, 105
Communication, 59–60, 86
Computers, 12, 35, 55, 57–87, 89, 96, 101–2, 104, 109, 117, 127–30
Constructivists, 31
Control, 43, 46, 48, 60, 69–70, 75, 79–80, 86, 91, 104
Le Corbusier, 13, 25, 32, 36, 38–39, 40, 52, 54, 100
Cubists, 31
Cybernetics, 83–87

Cyberpunk, 107, 109–12
Cyberspace, 62, 66–68, 76–88, 102
Cyborg, 84–87

Democracy, 50–51
Depression, Great, 50, 70, 82
Detournement, 97
Dewey, John, 5–10, 55
Dystopia, 98, 109–11, 115

Efficiency, 36, 39, 41–43, 51–53, 100, 112, 122, 124
Electrotech, 4, 16, 18–19, 28, 31, 57–88, 95–96, 99, 109, 112, 118, 122, 127–29; perceived as mechtech, 35

Factory, 2, 4, 19, 22, 25, 29, 35–36, 40, 44, 47–48, 50–51, 89–90, 94, 97, 99, 103, 107, 109
Fascism, 52, 90, 115–37
Firearms, 29, 35, 47, 102, 124, 127, 131, 134
Fragmentation, 93, 95, 98–100; specialization, 39–41, 44, 125–26
Function and functionalism, 14–18, 31–35, 39, 64, 73, 96–97, 99–100, 103
Futurists, 31, 36, 52

Gender. *See* Machine aesthetic categories, gender
Geocities, 80
Geometry, 38, 41, 50, 123, 134
Gibson, William, 109, 111–12
Gropius, Walter, 31, 54

Hackers, 62, 74, 76–79, 85, 102, 110
Hypertext, 80

Industrial Age, 11, 13, 29–30, 40, 49, 58
International Style, 73
Internet, 80, 82
Irony, 89, 93, 95, 99, 108, 129

Keyboards, 59, 75–76, 78

Locomotives, 12, 34–35, 58, 64, 73–74, 126

Machine, 10–11; decoration, 11, 30, 44–45, 73; loss of faith in, 93, 106–7; metaphor, 122–24, 129, 131; superiority over humans, 108–9; versus nature, 12, 45–46, 50, 96, 110
Machine aesthetic, 11–18, 26, 118; absence of, 7; mixed modes, 112–13; schemes of, 15–16, 18–22, 49
Machine aesthetic categories, 15, 18–22, 28; dimensionality, 18–19, 33–35, 44, 61–68, 95–96, 124, 127–28; dominant relationship, 18–19, 21, 46, 77, 102–3; erotic, 18–19, 21, 46–47, 77–79, 103–4; gender, 18–20, 43–45, 68–76, 100–101, 125, 128; motivating context, 18–19, 22, 47–48, 79–81, 104–6; object, 18–20, 37–39, 71–74, 97–99; persona, 18–21, 45–46, 76–77, 101–2; production, 18–20, 39–43, 74–75, 99–100; subject, 18–20, 35–37, 68–71, 96–97
Meaning, 3, 22–23, 25, 27–28, 101
Measurement, 38, 122
Mechtech, 4, 18–19, 28–55, 60, 63–64, 70–73, 79, 82, 87, 95–96, 99–100, 102, 112, 118, 122–27
Media, 47, 69–70

Metallic look, 38–39
Milwaukee, 1, 2
Missiles, 73, 81
Modern and modernism, 5, 17, 33, 54, 64, 73, 98, 110
Morality and ethics, 24, 53–54
Music, 36, 75, 89, 107

Neuromancer, 78, 111–12
Nietzsche, Friedrich, 23, 91, 103–4
Nostalgia, 96–97

Order, 41–43, 46, 48, 83, 91–92, 100, 104–5, 124

Patriarchy, 43, 68–72, 75
Perfection, 39–41, 44, 48, 75
Photography, 30
Plato, 52
Poetics, 23, 55
Polanyi, Michael, 62–63
Postmodern, 5, 33, 54, 90, 93, 95, 97–100, 107, 110–12
Power, 46, 74, 112, 124, 133–34
Progress, 49–50, 93, 112

Racism, 108
Rationalization, 37, 42, 53
Rhetoric, 2–3, 6–7, 22–28; defined, 22–23; of machine aesthetics, defined, 24, 28; technology and, 24–28; uses of in machine aesthetics, 30, 48–54, 81–87, 106–12
Rhythm, 36, 99, 125–26

Robot, 36, 47, 83–87, 90, 104, 127
Romanticism, 11, 30, 91

Science fiction, 89–90, 109–10
Signs, 2, 22–23, 60, 71, 111, 117
Simplicity, 41, 45
Simulation, 66, 69, 78, 82, 100–101, 109, 113–37
Specialization (fragmentation), 39–41, 44, 125–26
Speed, 39
Streamlining, 17, 68, 72–74, 83
Sullivan, Louis, 14, 30

Technique, 42, 53
Technology. *See* Machine
Telephone, ix, 57–59, 65, 71, 127–28
Television, 59, 70, 79, 81, 89–90, 94, 115, 117, 119–20, 127, 129, 134
Terrorism, 116–18, 120, 129, 134–35

Uniformity, 41, 126
Urban, 94–95, 98–99, 105, 107–8, 110–12, 123
Utopia, 51–53, 76, 81–83, 112

Violence, 46, 124–25, 134–35

Wiener, Norbert, 83
Wright, Frank Lloyd, 30

About the Author

BARRY BRUMMETT is Professor of Communication at the University of Wisconsin–Milwaukee. He is the author of several books and articles, including *Contemporary Apocalyptic Rhetoric* (Praeger, 1991).